Suit is a Four Letter Word

A Geotechnical Engineer's Introduction to Professional Liability

By
Hugh W. Nasmith, P.Eng.

With Contributions by:
Lech S. Brzezinski, P.Eng.
John W. Gadsby, P.Eng.
Bryan S. Shapiro, L.L.B.

Published by:

BiTech Publishers Ltd.
Suite 801 - 1030 West Georgia Street
Vancouver, British Columbia
Canada V6E 2Y3

ISBN 0-921 09500 7

PRINTED AND BOUND IN CANADA

Preface

The practice of geotechnical engineering has become a high-risk business. The incidence of professional liability claims in the U.S.A. and Canada has increased. Amounts claimed in suits are disproportionate to consultants' project fee revenue. This unhealthy climate motivated us to consider writing a text with case histories illustrating important aspects of professional liability in geotechnical practice.

Our paths first crossed during graduate engineering studies at the University of Illinois in 1960. The lectures and philosophy of Dr. Ralph B. Peck alerted us to the unknowns and the value of judgement in geotechnical engineering. We went on to develop consulting careers independently in eastern and western Canada and would exchange views on our engineering experience whenever we met. A common element in our work was a growing involvement with the examination of liability claims based on alleged errors or omissions in reports of subsurface conditions or in geotechnical recommendations.

In 1982, following a workshop of the Canadian Geotechnical Society, we realized that there was no available text to guide geotechnical engineers on management against exposure to claims.

Through our respective experiences with actual claims we had learned valuable lessons which might form the basis of a text. The only problem was that neither of us had time to write a book. Here we were, two "Characters in Search of an Author."

In 1982, Hugh Nasmith announced his retirement from his primary consulting activities. As president of a western Canadian geotechnical company, he had become increasingly aware of the interdependence between communication skills and good business practices in reducing the risk of professional liability claims.

We had found our author.

As plans for a book began to crystallize, we enlisted Bryan Shapiro into our team. Bryan is a lawyer who specializes in construction law, insurance and contracts.

It was agreed that the book should discuss law, insurance, business and reporting practices, the importance of clear, documented communications and the pitfalls which result in needless liability exposure. Case history examples were considered a key element to emphasize typical situations in practice, as well as risks and responsibilities to be avoided.

For the practicing geotechnical engineer, the consequences of today's decisions may not be fully realized until some years into the future. However, experience is the glass through which one may contemplate a surer path over rocky ground.

The purpose of this book is to provoke thought and discussion which will generate improvements in the training and professional development of geotechnical engineers. We emphasize the need for pooling information on professional practice claims so that all may derive the benefits of shared experience.

Vancouver, B.C. John W. Gadsby

Montreal, Quebec Lech S. Brzezinski

Canada

May 1986

Acknowledgements

The author has benefitted from the advice and assistance of his colleagues Lech Brzezinski, P.Eng., John Gadsby, P.Eng. and Bryan Shapiro, L.L.B. The draft manuscript was reviewed by Charles F. Ripley, P.Eng., and G. N. Kent, P.Eng., Ken Pepper, V.P. of Claims of Encon Underwriting Managers and National Program Administrator, and David V. Pym, President of International Specialized Risk Management (a member of the Simcoe & Erie Group of companies), participated in a number of valuable discussions of insurance problems. Chapter 3 is based on notes from a lecture of Bryan Nakai, P.Eng. and the pen and ink sketches were prepared by Alan Cameron.

The author acknowledges the contributions of a multitude of clients over a period of more than twenty five years in the geotechnical consulting business. The small minority of "difficult" clients provided much of the experience on which the book is based and the vast majority of clients have enabled the author to retain his faith in the honesty and fairness of the people with whom geotechnical consultants do business.

Hugh W. Nasmith

Victoria, B.C., Canada

April, 1986

Contents

Suit is a Four Letter Word

A Geotechnical Engineer's Introduction to Professional Liability

1 Introduction

"WHERE SHALL I BEGIN, PLEASE YOUR MAJESTY" asked the white rabbit. "BEGIN AT THE BEGINNING," said the King, very gravely, "AND GO ON TILL YOU COME TO THE END, THEN STOP."

Alice in Wonderland

Grounds for a claim against the geotechnical engineer may develop during construction or following completion of a project. Claims developed during construction are generally based on the contention that some subsurface condition was encountered that was not anticipated, or because there was greater difficulty in executing the below ground work than contemplated, or more of something than contemplated. Construction claims based on subsurface conditions usually result where there are no clauses in the contract documents that would permit an equitable adjustment to be made. Such claims generally attack the factual accuracy of the geotechnical engineer's report.

Unsatisfactory performance of a completed structure may also result in a claim situation against the geotechnical engineer. Unfortunately, in practice, clear and simple soil conditions are rather uncommon and the geotechnical engineer must properly understand the geological processes resulting in deposition and formation of the deposits and the limitations and deficiencies of the different techniques used to explore, sample and test the strata. Unless he does so, sophisticated analyses will be of no avail and predicted performance may be substantially in error.

For many geotechnical engineers beginning to understand their liability as Professionals comes when they are faced with a claim for an error which they are said to have made. Reading this book may give them a head start on this learning process.

This book is addressed to the young geotechnical engineer starting to work for a consulting firm and to the intermediate or senior engineer who is responsible for supervising him. Geotechnical engineers who have been working in the consulting field for many years will probably find nothing new in these chapters though it may be comforting to realize that your experience is not unique. The book may be useful to others (architects, structural engineers, managers, etc.) who from time to time are obliged to seek the advice of, or read reports prepared by, geotechnical consultants.

The objective of the book is to help the geotechnical engineer avoid claims and, if involved in a dispute, to minimize his exposure. However, even if you read the book ten times and faithfully follow all its advice you cannot be assured of a safe passage through the minefields of the consulting business. The surest way to avoid a professional liability suit is to go into some other field of endeavour, although even this is not infallible. By completing your studies to become a professional engineer you have identified yourself as an expert and even an innocent offhand comment may be construed by the courts as professional advice.

Professional liability cases are like aircraft accidents in that for every disaster there are probably a hundred "near misses".

The assumption is made throughout the book that the engineer is technically competent and will behave in an ethical and professional way toward his client. The world being what it is, the ranks of geotechnical engineers contain some incompetent and some unethical engineers. Nothing in the book is intended to assist these "bad apples" in avoiding the legitimate consequences of their behaviour.

The book consists of ten case histories interspersed with chapters which discuss a variety of topics related to the problems of professional liability and the risk of being sued. The case histories are chosen to illustrate important aspects of professional liability. The intervening chapters do not relate directly to the case histories which precede or follow but address certain principles in a general fashion.

The selection and presentation of case histories causes some difficulties since the author has no wish to embarrass his colleagues in the geotechnical field. **Professional liability cases are like aircraft accidents in that for every disaster there are probably a hundred "near misses".** The disasters are investigated from every possible angle. In the case of professional liability cases they are subjected to scrutiny by the courts. As guides for loss prevention the "near misses" are more valuable. However, consultants like airlines are reluctant to discuss let alone publicize their near misses.

The case histories in this book are of two kinds. Some of the cases are derived from court records, judgements, transcripts of evidence and documents submitted to the court. These have been written in a narrative fashion so as not to identify by name the people and organizations involved.

Other case histories are fictional accounts of professional liability situations. They are fictional, in that they are not based on a specific case but are derived from observation of the whole professional liability scene. Although they represent real life situations any resemblance to specific events and people is purely coincidental.

The book discusses law, insurance and engineering but is not intended to be a text book on these subjects. In general it is not a lack of knowledge of law or skill in engineering that leads to claims. Rather **failures in sound business management and communication are the ultimate cause of most claims and it is in these areas that improvements must be made to reduce the risk of lawsuits for professional liability.**

During the past twenty-five years significant changes have occurred in the field of professional liability for geotechnical consultants in Canada and the U.S.A. There have been significant advances in the analytical methods of geotechnical engineering which, for difficult sites or certain problems may give an appearance of precision which is not justified. Clients are much more knowledgeable and their expectations are higher. Society in general has become more inclined to litigation and responsibilities imposed on professionals by the courts have become more onerous. There is little likelihood of these trends being reversed. It is essential for all geotechnical engineers and especially those in the consulting business to take the steps necessary to reduce their exposure to professional liability suits.

Recommendations contained in this book involve some extra effort and occasionally may result in refusing an opportunity to work for a client. The reader may wonder whether the effort is worthwhile. In a healthy economic environment a consulting firm should expect to show a profit of from eight to fifteen percent on the annual fees. A hundred thousand dollar job should on the average produce a profit of ten thousand dollars.

Failures in sound business management and communication are the ultimate cause of most claims and it is in these areas that improvements must be made to reduce the risk of lawsuits for professional liability.

Losses to a consulting firm as a consequence of a professional liability claim include cash settlements not covered by insurance, together with the cost of non-chargeable work by senior personnel reviewing the file, negotiating a settlement or resisting a claim in court. These costs, in even minor disputes, seldom amount to less than ten thousand dollars and come directly from the profits of the firm.

Most consultants would spend significant business development effort in pursuit of a $100,000 assignment. Similar effort is justified in loss prevention measures.

Some idea of the magnitude of the problem is indicated by the fact that professional liability claim frequency in Canada has increased from one claim for each 20 design consultants in 1966 to one claim for every 3 firms since 1978. This does not reflect a decline in the standards of design consultants but instead is an indication of the changing environment in which they work.

Suspended rail line after shallow V-notch surficial slide in competent, very stiff to hard silty clay. Caused by run-off water ponded locally on uphill side of the track.

Did the geotechnical report recognize the importance of not impeding the natural downslope drainage when designing the side-hill / track cut and fill alignment?

2
Errors & Omissions

"We have left undone those things which we ought to have done, and we have done those things which we ought not to have done; and there is no health in us."
The Book of Common Prayers

The actions of geotechnical engineers which result in claims for professional liability fall into two broad categories, technical errors and administrative omissions.

Technical errors are such things as errors in calculation, leaving pertinent information off a drawing, failure to recognize the significance of field observations, and just plain forgetfulness. Every engineer and technician is subject to these errors. **No one is error-free but an adequate system of independent checks should catch the most flagrant technical errors.**

No one is error-free but an adequate system of independent checks should catch the most flagrant technical errors.

Administrative omissions include such things as failure to have a properly worded contract, failure to keep proper records, failure to communicate uncertainties to a client, assuming responsibilility which properly belongs to someone else and many others. Administrative omissions result from lack of experience; from undue optimism; and from pressures of time in dealing with clients. The hazards of administrative omissions are only anticipated by experienced engineers guided by a recognition that "what can go wrong will go wrong".

When a technical error is uncovered most engineers will immediately recognize and acknowledge it. A typical reaction is for the engineer to strike his forehead and say "Of course it should have been 60 mm instead of 6 mm". If he immediately acknowledges the error the engineer may accept more than his share of losses which may or may not result from a technical error.

On the other hand many engineers refuse to acknowledge the existence of an administrative omission. Failure to promptly document project arrangements and subsequent observations and instructions may lead to grief when something goes wrong. The reaction is commonly "I told the superintendent that he was cutting that bank too steeply, do I have to write a letter to the owner, architect and structural engineer as well?" As a result he may not recognize that he is involved in a potential claim and may not establish a defensive position until the situation is out of control.

Whenever information or advice is transferred from a consultant to a client, the consultant takes on a certain responsibility for the accuracy of the data and his recommendations. He also accepts responsibility for the consequences of the client's actions which are based on this advice. The view of the courts is that since the consultant has presented himself as an expert in his field the client is entitled to reasonably rely on his advice.

Neither the client nor the consultant foresees all the uses which may be made of the advice.

At this point of interchange of advice and responsibility, communication frequently breaks down. **Neither the client nor the consultant foresees all the uses which may be made of the advice.** The client does not understand the limitations inherent in the advice and the consultant does not recognize the extent of the responsibility which he is assuming.

The consultant becomes responsible for the foreseeable use which is made of his advice and if his report does not clearly define limitations in the use of the report, he may inadvertently become responsible for the misuse of his information and advice. Improper use of information and advice is a more frequent cause of claims than is inaccurate information or wrong advice.

3
Case History I

An incorrectly located borehole is probably one of the more common errors in geotechnical engineering. In most cases the error goes undetected since site preparation will destroy the evidence of the actual location of the borehole, and for most purposes a wrongly located borehole is not critical unless it actually happens to be on the wrong property. The following account describes a case where a wrongly located borehole was the basis of a complex dispute though in this case it appears that the losses suffered by the plaintiff arose from other causes than the consultant's error.

The client was an industrial firm which employed a firm of architects to design and supervise the construction of a building to house heavy machinery. The foundation investigation was carried out by a firm of geotechnical engineers who were familiar with local practice and conditions.

An incorrectly located borehole is probably one of the more common errors in geotechnical engineering.

The subsurface investigation was conducted in two stages. In the first stage the general site conditions were identified and it was recognized that the heavy machinery loads would have to be carried on end-bearing piles driven through soft clay to a very hard bedrock. From local experience it was known that the bedrock surface was very irregular and that difficulty in seating piles on steeply sloping surfaces was often encountered.

In the second stage a further drilling program was carried out to explore in more detail the area to be occupied by the building itself. Unfortunately three borings were plotted approximately fifty feet from their actual locations.

The drilling information from the two programs and reference to the geotechnical report was included on the bid documents along with a disclaimer which stated "This information is presented for the foundation sub-contractor. He shall satisfy himself as to prevailing conditions, and no extras will be allowed should conditions differ from those indicated."

The contract for the building was let to a local general contractor who called for proposals for the installation of end-bearing piles as shown on the bid drawings. Contrary to the recommendation of the local geotechnical engineer the client insisted that the contract for piles be on a lump sum basis.

Proposals were received from several contractors. The successful piling contractor had not worked in this area previously and proposed an alternative to the type of pile shown in the bid documents. Acceptance of the alternative required structural analysis and some modification of the piling system to achieve the same results as the system shown in the bid documents.

By this time all of the facts of the case had been explored and the only uncertainty remaining was how the judge would interpret the facts.

During the negotiations prior to the award of the contract the piling contractor carried out some drilling at his own expense and confirmed that bedrock slopes steeper than 45 degrees would be encountered. After the contract was signed but before any work had started a further revision to the pile system was proposed by the piling contractor and was accepted on the condition that there would be no change in the lump sum price for the piles.

The piling contractor was required by the terms of the contract to employ the geotechnical firm to provide inspection services. The geotechnical firm was thus fortunate to have an inspector on the site to obtain firsthand knowledge of the pile driving records. However, the piling contractor withheld payment for the inspection service and the geotechnical firm was never paid for this work.

Shortly after the first few piles had been installed the technician who was inspecting the work identified the error in borehole location and correctly surmised the cause of the error. The technician first advised the contractor on site of the error and then advised his employers the geotechnical engineering firm.

The piling contractor immediately entered a claim for extra piling although at this stage the total amount of piling which would be required was still unknown. Work was suspended briefly but the contractor agreed to continue and complete the work prior to resolution of the dispute.

Negotiation of the claim and threat of litigation continued for almost three years and was finally settled less than half an hour before the claim was to be heard in court. **By this time all of the facts of the case had been explored and the only uncertainty remaining was how the judge would interpret the facts.** The negotiated settlement involved a payment to the piling contractor of approximately 15 percent of the amount claimed as an extra.

The key question in the dispute was whether or not the contractor had actually suffered a loss as a result of the error in location of the boreholes. The dollar value of the contractor's claim was based on the number of feet of piling actually installed minus the number of feet of piling which he claimed to have estimated in making his bid, multiplied by the per foot allowance in the contract for piles added or deleted.

Although it might appear that the number of feet of piles actually installed would be easily and accurately determined even this figure was in dispute. The bid documents envisaged piling being installed from the bottom of the excavation for the basement of the building. The contractor actually installed piles from the original ground surface and cut them off when the basement was excavated. This amount of excavated and discarded piling actually exceeded the amount of the claim. The question thus arose as to whether the contractor originally envisaged installing piles from the surface or from the bottom of the excavation.

The contractor's case was weakened by the fact that he never documented the method by which he had arrived at the lump sum price for pile installation.

Since the geotechnical engineer had firsthand detailed knowledge of the length of each pile installed he was able to make a detailed comparison between the length of piles installed and what would reasonably be estimated using boreholes plotted in both their correct and their incorrect locations. Various methods were used including, drawing contours; assuming each borehole was representative of a proportional area; or simply averaging the borehole lengths and multiplying by the number of piles on the assumption that the boreholes statistically represented the topography of the bedrock.

Although an obvious and embarrassing error had occurred the geotechnical engineer did not automatically accept the claim that the contractor had suffered a loss as a consequence of the error.

The most detailed evaluation gave an estimate of total pile length which was closer to the actual amount when using the boreholes plotted in the wrong location than with the correctly located boreholes. The drilling program was never intended to accurately determine the footage of piling required and in fact was inadequate to achieve this result. This was the reason that the geotechnical engineer was opposed to the use of a lump sum piling contract.

It appears that the piling contractor suffered a loss as a consequence of putting in a low lump sum bid in a situation where he was not familiar with the local conditions and where he experienced a great deal of difficulty in installing the piles using his preferred methods. The appearance of the error gave him a fortuitous opportunity to attempt to recoup some of his losses.

It was never tested in court whether or not the disclaimer would limit the owner and architect's liability although the piling contractor entered a claim only against the general contractor and the geotechnical consultant. The general topic of disclaimers is discussed in Chapter 12.

This case illustrates the fact that claims for errors and omissions are sometimes used as matter of standard business practice. In this case **although an obvious and embarrassing error had occurred the geotechnical engineer did not automatically accept the claim that the contractor had suffered a loss as a consequence of the error.**)

Site investigations are commonly carried out by borings at intervals of about 30 m. Should this type of investigation be able to predict the in-situ conditions exposed by excavation on these two photos?

Very irregular saw-tooth bedrock profile.

Colluvial boulders mixed with soil.

4

Law and the Geotechnical Engineer

"THE STATUTE BOOKS ARE EXCEEDINGLY MUDDLED. I SELDOM LOOK INTO THEM" — Mathew B. Begbie — the hanging judge, later chief justice of British Columbia, before an open court, quoted by D.A. McGregor in "Sir Mathew Begbie"

Canadian Portraits, CBC broadcasts 1940, edited by R.G. Riddell.

Law is the custom of a community which is formally recognized as binding on the members of the community and enforced by an authority which rules the community. Law is written either as statutes or as guiding principles.

Canadian law is derived from the British system and consists of the principles of common law together with statutes or codes enacted by the Parliament of Canada or by Provincial or Municipal governments.

Common law is made and modified by the courts in response to the circumstances of specific cases brought for judgement.

Common law originated in medieaval England when judges appointed by the king travelled through the country to administer justice. Their decisions, which were based on common sense, were recorded in a central registry and these decisions and the facts on which they were based were used by the judges in making subsequent decisions. Thus, principles of law were developed which were common throughout the kingdom in contrast to feudal law which varied from place to place.

Common law was flexible since if a judge felt the circumstances of a particular case justified it, he could give a decision which was at variance to earlier decisions. If this judgement was upheld by the superior courts a new principle of common law was established and future judgements would be based on this new principle.

Thus the common law was flexible and capable of changing to meet new circumstances and changed standards of conduct. In effect, the **common law is made and modified by the courts in response to the circumstances of specific cases brought for judgement.**

The statutory law is a code enacted by a governing authority. The courts interpret the meaning of the code in cases brought for judgement and their interpretation establishes the meaning of the law. Not infrequently the court's interpretation differs from what the legislators intended and the legislature is then obliged to revise the wording of the statutes in order to achieve its objectives. **A statutory or codified law is less flexible than common law and the meaning of the statute is only established in law when it has been tested in the courts.**

In Canada such matters as criminal law, taxation, etc., are covered by statutes. However, many matters which give rise to disputes in the geotechnical field are covered by common law. For this reason judgements given in earlier cases, including cases before the courts in Great Britain, although not binding on Canadian judges, may influence the court and are frequently cited as precedent in Canadian cases.

A statutory or codified law is less flexible than common law and the meaning of the statute is only established in law when it has been tested in the courts.

Law is not fixed and unchanging. Both common law and statutory law are continuously changing in response to changes in the demands of society. The outcome of any dispute before the courts is not completely predictable and this, together with the high cost of litigation, encourages settlement of most disputes without resorting to the courts.

In order to bring a dispute before the courts for a decision the plaintiff will define a "cause of action" which may be described as a 'legal classificiation for a type of indiscretion for which the courts will provide a legal remedy'. In the following paragraphs some causes of action with geotechnical implications are discussed including ***Support, Trespass to Land, Private Nuisance, and Strict Liability.*** If a geotechnical engineer is involved in a case where the cause of action is one of these, the litigants will almost certainly endeavour to claim against the engineer on the basis of ***NEGLIGENCE*** which is also a cause of action.

SUPPORT - By common law an owner of land is entitled to lateral support for his land in its natural state. This is a right which goes with the land itself. Where the activity of a landowner causes movement of the adjoining property the owner of the adjoining property would have cause for action against the person who caused the movement. The failure of an excavation which causes damage to the adjoining property is an obvious case where failure to provide support would be the cause of action. Less obvious cases can occur where activity on one property triggers erosion and removal of soil from another property, not necessarily adjoining.

It has recently been established in Canada and the U.S. that an adjacent landowner has the right to be protected against interference with the support provided by groundwater below his land. Geotechnical consultants have been implicated in major lawsuits where temporary or permanent ground water lowering has caused adverse settlement damage to adjacent properties.

TRESPASS TO LAND - is defined as direct interference with land which is in the possession of another. There are only two significant defences to this cause of action, consent and involuntariness. If you have consent for actions which interfere with the land, you are protected against a claim for trespass. However if you have consent and do something different than what was envisaged or permitted by the consent, then you may be guilty of trespass. Intention to trespass is not required. A trespass in error is still trespass.

The case history described in Chapter 7 is an example of trespass in the geotechnical field. The plaintiff had an agreement with the defendent who was the owner and developer of an adjoining property to install anchors which extended under the plaintiff's building to provide ***temporary*** support. It was later discovered that the wall was underdesigned and the anchors were required for permanent support. Action was brought against the defendants including the engineers for trespass and the courts awarded damages.

PRIVATE NUISANCE is an indirect interference with private property. There is a subtle difference between this cause of action and that of trespass to land. Roughly speaking it is the difference between spinning your wheels so as to throw a rock which breaks a window in your neighbours house and entering your neighbours property and breaking a window with a hammer.

Law is not fixed and unchanging. Both common law and statutory law are continuously changing in response to changes in the demands of society.

STRICT LIABILITY may be a cause of action where the owner of the land brings something onto his land which constitutes a non-natural use of the land and it escapes and causes damage. The owner is liable for the damage. This is based on a classic case where the defendant built a water reservoir on his property. The plaintiff operated an underground coal mine. Ancient works connected with the plaintiffs mine and they were not detected by the defendant or his engineers. The water escaped into the plaintiffs mine and caused damage. The courts found in favour of the plaintiff stating that "anyone who for his own purposes brings on his land and collects and keeps there anything liable to do mischief if it escapes must keep it at his peril."

In an action for strict liability no fault needs to be shown. It is only necessary to establish the three elements, ***accumulation, non-natural use*** and ***escape causing damage.*** The only defences available are that it was caused by an act of a stranger or as an act of God. Strict liability has been applied in cases involving fire and explosions, spraying pesticides and herbicides, escape of water from a dam or dyke, vibrations from pile driving and escape of mine tailings from storage. Strict liability may also occur in situations where the parties have agreed by contract that the principle of strict liability shall apply.

NEGLIGENCE as a cause of action is a rapidly growing area of law. There are three elements which must be proved by the plaintiff in a case in which negligence is claimed. First the plaintiff must prove that the defendant owed a duty of care. Second the plaintiff must prove that the defendant breached that duty of care. Finally the plaintiff must prove that he suffered damages as a result of the defendant's breach of the duty of care.

For a professional engineer the duty of care is clear. He presents himself as having special skills and is hired to give advice. Any contractual arrangement brings about a duty of care. However the duty is not limited only to those with whom he has a contract.

An example of this is found in the following case. A project involved the demolition of an existing building but it was decided to leave a wall in place temporarily. The architect in charge asked the demolition foreman if the wall was safe, but did not examine it himself. The builder moved in a tool shed and the wall collapsed killing two workmen and injuring the plaintiff. The defendant architect argued that the plaintiff, the builder, was not a party to the demolition contract and that therefore there was no duty of care. The court held that the builder was in a class of people whose safety was the duty of the architect and therefore he owed a duty of care along with the demolition contractor.

Another situation where the question of duty of care arises is in the case of gratuitous information or advice. It has been established by the courts that if a person who has special skills as a professional engineer, gives gratuitous advice where he knows or ought to know that the person receiving the advice may act on his advice and the person reasonably relies on that advice, then there is a duty of care on the person offering the advice.

What this means is that any time you answer a question which relates in any way to your profession as an engineer, a duty of care will arise unless you provide an adequate disclaimer at the time you give the advice. This goes to the question of reasonable reliance. It is not reasonable to rely on advice which has been adequately disclaimed at the time it was given. However, professional engineers cannot go around arbitrarily affixing disclaimers to their advice. (The topic of disclaimers is discussed further in Chapter 12).

The accepted standard of care provides the test for what constitutes a breach of the duty of care. This standard varies for different situations and relationships. For professionals it is what a reasonable professional would have done at that time under similar circumstances. In practice therefore, you don't have to provide perfection or even the highest standard of professional advice as long as your advice is reasonable and conforms to current practice at the time it was given.

In professional liability litigation where the standard of care is a major issue both sides will usually bring expert witnesses to support their contentions. The outcome of the case can turn on which expert witness is more credible because the burden of proof is the balance of probabilities. It can be a simple matter of the more convincing expert witness tipping the scales in favour of one side or the other.

As the third element in the cause for action for negligence, the plaintiff must demonstrate that he suffered damages and that the damages were foreseeable and were the consequence of the defendent's negligence. The case history described in Chapter 3 is an example of a case where although the geotechnical engineer made an error, the losses suffered by the plaintiff were not the consequence of the engineer's error.

This has been a very brief summary of some of the legal principles which may affect the geotechnical engineer in the course of his professional work. From the point of view of errors and omissions liability, claims against him are most likely to be based on the tort of negligence. He may be attacked for negligence in carrying out his professional responsibilities in a contract with his client or negligence may be claimed in his duty to third persons with whom he has no contract but who nevertheless rely on his professional actions.

5 Case History II

This case history illustrates problems resulting from a report which gives an unrealistic impression of precision aggravated by the risks of a remote site.

The client, a mining company with worldwide interests employed a general engineering firm to design and supervise construction of a harbour facility. The general consultant employed a geotechnical firm to investigate subsurface conditions for a wharf which eventually was supported on a piled foundation. The geotechnical firm sent an engineer to the site to set up a drilling program. Due to delays in mobilizing equipment the engineer left the site before drilling began, and the program was carried out by a local driller under the direction of an experienced technician.

The offshore drilling program consisted of a number of wash borings with standard penetration tests attempted every 1.5 meters. A few dynamic cone penetration tests were made adjacent to some of the wash borings. The results of the drilling program were forwarded to the geotechnical engineer's office for analysis and reporting, and the technician was moved to another project with only a brief visit to home base.

The drilling program revealed a gravelly sand to a depth in excess of 50 meters. The "N" values from penetration tests indicated that the sand was in a compact to dense state. The uncased dynamic cone tests showed low blow counts increasing with depth as the friction on the rods increased and were terminated at a practical refusal depth of 15 m. Notes in the technician's field book indicated that there were problems with sand heaving in the casing between sampling intervals and in some instances samples could not be taken for this reason. Some attempts were made to use drilling mud to prevent heaving but this was not successful.

The geotechnical firm prepared an elegant report with numerous design charts and lengthy discussion of design procedures and recommendations for foundations. A computed pile length of 15 meters was given for the type of pile and design load chosen.

Pile length predictions were based principally on the standard penetration test results since they were sufficiently numerous to be analyzed statistically. The dynamic cone results were largely ignored in the assessment of soil density.

The significance of sand heaving was overlooked and the drilling inspector's field notes were not sufficiently detailed to show whether or not the tests were being taken above or below the bottom of the casing. The false refusal in the cone penetration tests was not properly understood.

The design for the wharf was finalized, a contract was let and construction commenced the following spring. The first piles drove easily with no sign of bearing resistance at the design elevation. After consultation, further driving and a pile load test, a satisfactory length was established at approximately double the original length. Since there was insufficient quantity of piles on the site, construction was delayed for a year and both the owner and the contractor claimed against the geotechnical firm for losses caused by the delay.

In this instance the field data was inadequate for the use which was made of it and an error of 100 percent in the predicted pile length was clearly unacceptable.

However, even with the best standard penetration and other field tests, how accurately can the required pile embedment be predicted in the absence of a pile loading test? Would plus or minus 25 percent be reasonable. Would this be acceptable to a contractor who might be left with 25 percent or more of his piles left unused on the site? Could the uncertainty be communicated in the geotechnical report and incorporated in the contract documents so that the geotechnical engineer would not be subject to a professional liability claim.

The geotechnical engineer must keep in mind the ultimate uses which may be made of his report and the consequences of any errors or uncertainty which may be present in his data or design interpretations.

6

Anatomy of a Dispute

"A poor fellow consulted a famous lawyer about a parking ticket, and was advised to fight it in court; although the fees would not be cheap, it was a case which ought to be fought. They went to court and lost, despite extensive preparation and brilliant argument by the famous lawyer, who recommended an appeal; although the fees would be very heavy, there was an important principle at stake. The fellow mortgaged his house and paid for the appeal, which was lost. The lawyer recommended a further appeal citing all the errors in law which had been made by the judge, and this appeal too was taken, and lost.

Although the fellow was destitute, he agreed with his lawyer's strong advice to borrow heavily from his family and friends to finance a final appeal to the highest court in the land where, the famous lawyer assured him, victory had finally to come. In the highest court in the land the famous lawyer argued brilliantly, and lost. He wired his client the bad news. The client wired back: 'I took your advice and have lost everything —house, money, friends, family. What do I do now?'. The famous lawyer received the wire at his five-star hotel after leaving the highest court in the land, and immediately wired back: 'Go home and breed. The law needs men like you.'

from A Writer's Notebook by Richard J. Needham in the The Globe and Mail September 24, 1984.

The seeds of a dispute are often sown when the geotechnical consultant agrees to undertake a study or project for a client. With the benefit of hindsight it is often possible to see how the dispute might have been avoided if this initial contact had been carried out differently.

One of the common causes of disputes occurs when the consultant either does or fails to do something as a result of which the client's expectations are not met. For example, the engineer may make an error which costs the client extra money or the engineer may fail to warn the client that he may encounter unexpected conditions which will cost the client more money. The fault is seldom entirely with the geotechnical engineer. Frequently the client has unrealistic expectations as to what he will get from his consultant.

At some stage the client will express his dissatisfaction with the geotechnical engineer. Either verbally or in writing he will indicate that he expects the engineer to pay for some of his extra costs and threaten legal action.

As soon as the engineer is aware of any circumstances which may give rise to a professional liability claim whether he has been notified in writing or not, he must advise his professional liability insurer and provide a brief outline of the circumstances. If he does not do this and does nothing or if he takes significant action, particularly in admitting liability, he may invalidate his professional liability coverage.

As soon as the engineer is aware of a threat of a professional liability claim against himself, whether he has been notified in writing or not, he must advise his professional liability insurer and provide a brief outline of the circumstances.

If the risk of a professional liability claim seems relatively small and no claim has been filed in court the insurer may simply open a file and ask to be kept advised of developments. If, however, the risk appears serious, the insurer will appoint either a legal firm or a firm specializing in professional liability negotiations to review the circumstances of the claim and to act on behalf of both the insurer and the engineer. During this investigation stage the insurer's representative will ask for a copy of all significant documents in the engineer's files. From this point onwards the engineer must then clear with the insurer's representatives all correspondence and negotiations with the claimant.

As the dispute proceeds into the courts copies of all documents must be made available to the plaintiffs representatives. Since files and documents are lost from time to time, the engineer should endeavour to retain original documents in his files and take special care to ensure that they are not mislaid in his filing system.

From this point on, the management of the professional liability claim tends more and more to pass out of the control of the geotechnical engineer although he will continue to have to invest a significant amount of time and effort in the negotiations and litigation leading to the final settlement of the claim. He will do this to minimize his losses on the deductible portion of his insurance and as a matter of principle to protect his reputation as a competent geotechnical engineer. As well, he is obliged by the terms of his professional liability policy to cooperate with the insurer and the legal counsel appointed to defend the claim.

If the claim cannot be settled by negotiation, the plaintiff, whether he is the client or a third party, must employ a legal firm to issue a WRIT OF SUMMONS. This legal document identifies the plaintiff and the defendent and briefly states the basis of the claim. The defendent must respond by a stated time, otherwise the claim will be heard and settled in his absence.

The original document is served on the defendent (the geotechnical engineer), who will pass it on to his own legal counsel or the insurers representative for a response. If the insurer has not already done so, it will appoint a legal firm to act and from then on the management of the professional liability claim is directed by the insurer or its representative who with the legal firm will take appropriate action.

The first action of the lawyer may be to request a postponement for responding to the writ of summons or statement of claim. The legal firm will then review the facts of the claim and the legal position of the plaintiff and defendent, including an assessment of the damages which the plaintiff claims to have suffered. At this stage the insurer will often engage a well known outside consultant to study the technical facts in the case and later, if necessary, to act as an expert witness. Together with the insurer and lawyer they will endeavour to assess the risk of an adverse judgement in court and the possibility of a successful defence, and based on their assessment, advise the insurer and the geotechnical engineer of the best course of action.

Because of the high cost of litigation the next step is often an attempt to negotiate a settlement through the plaintiff's legal representative. If this is not successful, the plaintiff's lawyers will demand that the engineer's lawyers file a STATEMENT OF

DEFENCE which is a legal document setting out in legal language the basis of the defence.

The filing of the statement of defence is the first positive step on the road to the courtroom, but the path is neither short nor the destination inevitable.

The filing of the statement of defence is the first positive step on the road to the courtroom, but the path is neither short nor the destination inevitable.

The lawyer for the insurer and engineer will draw up a STATEMENT OF DEFENCE in which, with the assistance of the engineer, he responds in legal language to the claim filed by the plaintiff. Both of these statements are general in nature and represent a preliminary position of the adversaries in the dispute. Both the claim and the defence may be substantially revised as the dispute proceeds.

In due course, a date for the courtroom trial will be set and under normal circumstances this date will be set back at least once and possibly several times. With a date set, the preparations for a courtroom appearance speed up from a snail's pace to a slow walking pace. Because of the high cost of litigation many counsel defer action (employing expert witnesses, assembling documentation, etc.) as long as possible since there is always a good chance that the dispute will be settled by negotiation or that the plaintiff will abandon his claim. Proper preparation, however, means early preparation.

Before the case comes before the court EXAMINATIONS FOR DISCOVERY will usually be held. At this stage witnesses will be examined under oath, verbatim transcripts prepared, documents presented and reports by expert witnesses submitted. The purpose of the examinations for discovery is to permit each side to examine the detailed evidence of the opposing side, and to avoid trial by ambush. **Contrary to the television viewers impression of courtroom procedures, surprise witnesses and surprise evidence are not tolerated.**

After the examinations for discovery each side has a clear picture of the factual evidence available to the other side and frequently a negotiated settlement is reached. Less frequently, the plaintiff may abandon his claim.

If a negotiated settlement is reached the plaintiff, as part of the settlement, will submit a consent order to the court which consents to the dismissal of the plaintiff's claim against the defendant engineer. In addition a release will usually be executed in which the plaintiff agrees to release the engineer once and for all, from all claims arising out of the facts in issue in the current dispute.

A recent development in Canada in some jurisdictions is the mini- trial. The opposing lawyers present their cases before a judge who gives his non-binding opinion as to what judgement would be given at trial if the cases were presented with a full array of witnesses, cross examination, documentation, etc. A mini-trial should take ten to twenty percent of the time (and cost) of a formal trial. The judge's opinion in a mini-trial, although not binding, often forms the basis for a negotiated settlement between the disputants.

Contrary to the television viewers impression of courtroom procedures, surprise witnesses and surprise evidence are not tolerated.

If no settlement is reached, ultimately (anywhere from one to three years) the two disputants and their lawyers, and their witnesses, and their experts and their documents arrive before a judge and present their evidence and legal arguments.

In Canada, in contrast to the U.S. and other jurisdictions professional liability cases are not normally presented to a jury, for consideration, but only to a judge.

In due course, the judge hands down his written judgement. The disputants may accept the judgement or one or other may appeal his decision or part of it (e.g., the amount of damages) to a higher court.

The geotechnical engineer having appeared in court, and if required, paid his deductible portion of the insurance coverage is glad to put the experience behind him.

By now the whole business is entirely in the hands of the insurance company and the lawyers, provided of course the consultant had professional liability insurance. If he did not, he will have had to bear the entire legal cost of the investigation and litigation process himself, win or lose.

The geotechnical engineer having appeared in court, and if required, paid his deductible portion of the insurance coverage is glad to put the experience behind him.

Footnote

The discussion in this chapter is based on experience in British Columbia. Procedures may differ in other jurisdictions.

7 Case History III

An Engineer may be legally responsible for trespass if his design involves foreseeable trespass or nuisance.

This case illustrates that a subconsultant engineer, whether geotechnical or structural, may have a legal responsibility beyond the contractual responsibility to his client for technical work. The case illustrates some of the legal concepts of trespass and nuisance.

The developer proposed to build a 20 storey office and commercial complex on a site adjacent to an existing 14 storey building. The owner of the 14 storey building (hereafter referred to as the neighbour) through his lawyer expressed concern regarding the risk of damage to his building and notified the developer that he would be held responsible for any damage.

The developer employed a construction management firm (actually a subsidiary of the developer) and a firm of structural engineers to develop the engineering design and they in turn employed a geotechnical consultant to advise them.

The principal geotechnical problem was support of the neighbour's building during the excavation of the foundation and basement of the developer's building. The level of the excavation would be some fifteen feet below the foundation of the neighbour's building.

The geotechnical firm recommended stabilizing the excavation and preventing any lateral movement of ground under the neighbour's building by means of prestressed anchors into the wall of the excavation and extending 24 feet beneath the foundations of the neighbour's building. The pertinent portion of the geotechnical report is as follows.

"We recommend that these anchors be used only for temporary support and therefore the basement wall of the new structure must be capable of carrying the same loads.

In order to eliminate any danger involved with future excavation of the anchors following demolition of the adjacent building we suggest that the underground section of the tie-back anchor be grouted after the post tensioning has been reduced to 80% of design load."

The structural firm prepared a plan for the tie-back anchor-bolt system, a copy of which was sent to the owner of the neighbouring building. This drawing included a note which stated "Tie back anchors are required during construction only".

As ultimately constructed, however, the basement wall was not designed to carry the loads carried by the anchors so that the anchors were required to function after construction was completed.

As the construction proceeded the neighbour called in independent consultants and on the basis of their advice entered a claim for damages as a result of loss of support of his ground resulting from the continued presence of the anchor rods beneath his property. The claim was against the structural engineers and developer for negligence, for continuing nuisance and trespass.

The legal proceedings continued for at least three years. The judge in the lower court had to wade through the testimony of numerous expert witnesses and masses of technical data. He had to resolve two principal questions.

1. Were the anchor rods installed and left in place with the approval of the plaintiff (neighbour)? The judge concluded that although the neighbour had received a copy of the tieback/anchor plan and seen the bolts being installed he was justified in believing it was a temporary installation. The meaning of temporary is somewhat vague - temporary physical presence or temporary as to function. In any case the judge concluded that the anchor bolts represented a continuing trespass on the neighbour's property.

2. Had the neighbour suffered any damage or loss as a consequence of the continuing presence of the bolts beneath his property?

Based on a variety of conflicting and expert opinions the judge concluded that the anchor bolts would continue to satisfactorily support the neighbour's foundation for the anticipated life of the building and that if the neighour's building were demolished the anchor rods would present no significant hindrance to making a deep excavation. However the judge did accept the idea that some potential purchaser of the building in the future would be alarmed by the presence of the anchors and the owner might consequently be obliged to sell the property at a reduced price. On this basis the judge awarded a substantial sum to the plaintiff. This award was overturned on appeal to the Court of Appeal who however awarded damages of $10,000, an amount equal to a very small portion of the lawyers fees incurred in the litigation.

The engineers were made liable for the damages along with the developer because by preparing the plan of anchors and bolts they directed that they be installed in what constituted a trespass.

This case illustrates that the engineer must take responsibility for the consequences of his design beyond his contractual responsibility to his client.

8
Contracts

"A verbal contract isn't worth the paper it's written on" - Sam Goldwyn.

In spite of Goldwyn's low opinion of verbal contracts an alarming amount of geotechnical work is done on the basis of a telephone call. The fact that it is a verbal contract does not make it any the less binding, but in the absence of written evidence, serious problems can arise in attempting to determine what the exact terms of the contract were. If a professional liability claim arises from work undertaken on the basis of a verbal agreement, the courts will determine whether or not a contract exists and if it does, what are its terms. The court's decison will not always be to the satisfaction of the geotechnical engineer.

The first rule of contracts for the geotechnical engineer is "get it in writing".

The first rule of contracts for the geotechnical engineer therefore **is "get it in writing."**

A legally binding contract contains a number of elements which are defined or inferred whether the contract is a verbal agreement, an informal letter or a formal document drawn up by a lawyer.

A contract is based on an offer made by one party to the contract and accepted by the other. The geotechnical engineer or firm offers to provide specialized engineering services and if the contract is completed the client accepts the offer.

The contract must show the intention of both parties to enter into the contract. This is usually achieved by having copies of the contract signed by both participants. However, it is still a binding contract if the engineer offers his services in a letter and the client phones back and says "go ahead". Problems can arise however, when the the client says "go ahead, but –" and proceeds to change the terms of the offer outlined in the letter. If the engineer accepts these changed terms he should confirm them in a further letter and have the client acknowledge the changes.

A contract defines the consideration, that is it states what is to be provided by each of the parties to the contract. In the case of a contract for geotechnical services it describes the services which are to be provided by the engineer and the basis and method of payment for these services by the client.

The description of the services to be provided is often the weakest part of contracts for geotechnical services.

The description of the services to be provided is the weakest part of contracts for geotechnical services. If the description is vague or ambiguous and a dispute arises, the courts may be obliged to infer what the geotechnical engineer undertook to do, and this may be quite different from what the engineer had in mind when the contract was made.

Finally, ***to be legally binding a contract must be made between parties with the authority to enter into the contract.*** In most cases the offer to provide geotechnical services is contained in a proposal on the letterhead of the geotechnical firm and signed by a responsible member of the firm. Geotechnical engineers should endeavour to get authorization from the client in a form that clearly indicates who is responsible for paying their bill. If a contract is signed "for or on behalf of XYZ Company" it is the firm who will be responsible under the contract. This will also define who the geotechnical engineer is responsible to "in contract" though he will also be responsible "in tort" to third parties who are not named in the contract.

In summary then, a contract is based on an offer made by one party and accepted by the other. Both parties must be on an equal level of sophistication. And the contract must define what is to be done by each party to the contract.

Contracts may contain many other terms dealing with such topics as insurance, scheduling, access, ownership of documents, confidentiality, etc., which are intended to serve specific purposes and circumstances.

In addition to the verbal contract which should be avoided even if it means losing a client, there are three main types of contract that the geotechnical engineer will encounter. These are letter contracts, client-drawn contracts and standard contracts.

Letter Contract

The most common form of contract for geotechnical services is a letter on the consultant's letterhead addressed to the client. The letter offers to provide geotechnical services which are described and defines the basis of payment for these services. The letter is signed by the geotechnical engineer and contains a request that if the client accepts the proposal he should sign and return a copy of the letter. The wording of the acceptance might be as follows:

"Accepted and agreed to (date) (client) (position)".

Such a proposal letter is usually written by the geotechnical engineer and is based on previous discussions with the client, site visits and examination of drawings or other documents. These letters should be carefully written and reviewed by an experienced senior engineer in the geotechnical firm since the contents and wording will form part of the contract and will be significant if a professional liability claim should subsequently develop.

From the point of view of professional liability, a description of the purpose and the scope of the geotechnical work is the most important part of this proposal/contract. The proposed project, as it is understood at the time the letter is written, should be described in sufficient detail that significant changes made subsequently can be identified. In this regard any documents and drawings which were examined should be mentioned. As an example, an investigation which was suitable for a two or three storey flexible frame structure might be judged as negligently inadequate if it was used to design the foundations for a 12-storey reinforced concrete structure.

The objective of the investigation should be described as far as possible in specific rather than general terms. "A preliminary subsurface investigation" means different things to different people and at different times. However, "the investigation will provide information on which to base a decision regarding the appropriate type of foundation, piles, spread footings, etc." clearly indicates what type of information is expected from the study.

The proposal should indicate the scope of the work. In describing this, the engineer will have to strike a balance between being too specific and too vague. He might undertake to put down x number of boreholes, collect y number of samples and do z number of tests. Once the investigation has started the subsurface conditions may indicate a different approach and if the work done is significantly different from that proposed (even though the cost remains within budget) he may in effect be in breach of contract unless he has described the scope of work as conditional on subsurface conditions actually encountered.

The objective of the investigation should be described as far as possible in specific rather than general terms.

An experienced geotechnical engineer should have some idea of the amount of uncertainty regarding a subsurface investigation at a specific site. He should try to communicate some of this uncertainty to his client so as to avoid raising unrealistic expectations in the mind of the client. However, he is faced with the practical problem that if he communicates too much uncertainty the client may lose confidence in him.

The alternative of describing the scope of the subsurface investigation in vague terms is generally less desireable. The courts in considering a proposal/contract which contains a vague description of the work to be done will infer that the geotechnical engineer undertook to do an investigation meeting the current standards of the profession. The standard of the profession may be defined in court by the testimony of an expert witness hired by the plaintiff. He may say "I always drill at least six test holes" and decide that your three test holes was negligent. If on the other hand your proposal says, "we will drill three holes which we believe will give an adequate picture of subsurface conditions. This information will be suitable for the selection of the foundation type and capital cost estimates. Further drilling may be required for design purposes". The court will be more inclined to recognize that your drilling program was based on an engineering assessment of the site requirements.

If a proposal/contract is limited by a budget imposed by the client it is essential that this fact be referred to in the letter. Any reservation you have about the adequacy of the program under budget constraints must be voiced in writing and any report subsequently written must also indicate all factors which may affect its validity.

The proposal/contract is frequently written by the geotechnical engineer following discussions with the client at a time when the appointment of the engineer is assured. The letter is then essentially a confirmation of a verbal agreement. Under these circumstances the letter is often written without much care or review. Such a letter should always be written carefully and the wording reviewed. **Even the most straightforward job with the best client can go sour and a carefully written proposal/contract can be a valuable shield in an errors and omissions dispute.**

Client-Drawn Contracts

Many clients with a legal staff have a standard contract form which is used either directly or in a modified form for contracts with a geotechnical consultant. These contracts present a different type of problem for the geotechnical engineer.

Even the most straightforward job with the best client can go sour and a carefully written proposal/contract can be a valuable shield in an errors and omissions dispute.

Purchase orders are a special form of client-drawn contract used by many large organizations and industrial clients. These purchase orders are designed for the benefit of the accounting department for purchase of goods and routine services and may require modification for the purchase of specialized geotechnical services.

Many client-drawn contracts contain terms which are inappropriate for geotechnical consulting services. When this type of client-drawn contract is presented to the consultant it should be carefully reviewed by a senior engineer in the firm, and if necessary referred to the consultant's legal counsel for interpretation of the implications of the terms. Occasionally they contain terms such as guarantees or hold harmless clauses which would fall under the exclusion to the coverage provided by the consultant's professional liability insurance.

Following is a typical unacceptable hold harmless clause which would invalidate the engineer's insurance: —"The Engineer shall indemnify and hold harmless the Client from and against any and all claims, demands, actions, causes of action, suits, proceedings, authorities and Judgements of every nature and description brought or recovered against the Client, the Engineer, or both, by reason of any act or omission of the Engineer or the Contractor on the Project, their agents, employees, sub-contractor or sub-consultants in relation to the performance of their Work or Services."

Almost always client-drawn contracts are biased strongly so as to protect the client at the expense of the consultant. Most clients are willing to modify or delete inappropriate terms when the request is justified. If the client is inflexible, the consultant is faced with a practical business decision whether or not to enter into a contract which he hopes will not be enforced to his detriment.

Almost always client-drawn contracts are biased strongly so as to protect the client at the expense of the consultant.

Client-drawn contracts are often used when competitive proposals call for the provision of geotechnical services. When the contract is awarded the client will prepare a contract which consists of his standard contract using the consultant proposal as the description of the work to be done under the contract. The consultant's proposal is probably more carefully written and reviewed than routine letter proposals. However, the geotechnical firm has to strike a delicate balance in the wording of its proposal between the risk of failing to get the contract and the risk of being legally bound by the wording of an optimistic sales pitch.

Standard Contracts

The Association of Consulting Engineers of Canada has prepared recommended contract forms for the provision of geotechnical engineering services. Other organizations representing consultants have drawn up standard contract forms and some large consulting firms have had standard contracts drawn by their legal advisors. The contracts take into account the special services provided by architects and engineers and endeavour to achieve a fair division of responsibility and risk between the consultant and his client.

These standard contracts provide a useful check list of the topics of concern and can be used for comparison with a client-drawn contract. Because these contracts endeavour to cover every possible situation they are long and are almost never used by geotechnical engineers. Most clients of geotechnical consultants would be uneasy when presented with a long detailed contract in formal legalese and would be unwilling to sign such a contract without having it referred to their legal advisors.

Mephisto's UNETHICAL ADVICE

A regular feature of the Devil's Advocate and the Hades Daily News

Dear Mephisto:

I am a young lawyer and most of my experience is with transferring property and drawing wills. However, my brother-in-law has a contracting firm and has some trouble with a job he is doing. He hired our cousin to dig some pits with a backhoe to install concrete footings on bedrock. While he was doing this it rained for a week and the holes filled up with water, and while he was trying to rent a pump the finance company re-possessed the backhoe and the upshot was that my brother-in-law had to hire someone else to do the job. This cost my brother-in-law about $10,000. Should he sue his cousin?

L. Beagle III

Dear Boy:

Sue his cousin? Don't be silly - from what you've told me he's probably bankrupt and it's most unlikely that he has insurance - besides, blood is thicker than water isn't it? Go after the professionals, engineer and architects - they have the "deep pockets" as we say in the trade.

The first thing to do is get a copy of the soils report and study it carefully. If it says there is soft clay on the site you claim it doesn't say how soft. If it says soft clay in one place and firm clay in another say your client was misled. In any case use your imagination. Consider hiring another geotechnical engineer to review the project. It will be money well spent. He may have more imagination than you. If you don't like his report you don't have to use it.

Don't limit yourself to suing the soils engineer. Enter a claim against the architect, the concrete supplier, the building inspector and the engineer who designed the air conditioning. If they each contribute $2000 you've got your $10,000 right there. But why limit yourself to $10,000. There are delays and interruption of schedules, etc.

Yours for big and imaginative thinking

Mephisto

9
Case History IV

An Engineer's liability extends beyond his client to the ultimate owner.

A geotechnical engineer may feel that when he is employed as a sub-consultant on a job the prime consultant is responsible for properly using his advice and where necessary passing that advice on to the client. The following case illustrates the fallacy of this assumption.

The client in this case was an elected public body who was responsible for providing a building to house staff and facilities to serve the community.

The prime consultant was an individual architect who had worked closely with the client over a period of years and was a member of the committee responsible for selecting a site for the building. The architect had designed and supervised the construction of other buildings for the client near the proposed new building. The architect had a standard contract with the client to provide services and supervision of construction for the proposed building.

The sub-consultant was a small firm of consulting engineers engaged principally in the field of structural engineering. From time to time they had worked with the architect to provide minimum structural design which the architect incorporated in his drawings.

The architect arranged with the client to provide a backhoe to dig two test pits on the site and called the structural engineer to send someone out to inspect the test pits. The structural firm employed, on a part time basis, a retired engineer whose expertise was principally in structural design but who had some practical experience with soils though he made no claim to be a geotechnical engineer.

This employee of the structural firm met the architect on the site and inspected the test pits. He correctly described the soil that was exposed in the test pits and suggested a bearing capacity for the soil. He also told the architect that he felt that deep borings and a proper geotechnical evaluation was required. The architect replied that the client would not be prepared to pay for such an investigation.

The employee reported his observations and comments to one of the principals of the structural firm who phoned the architect and was again told that the client would not pay for deep borings and a proper geotechnical study. The structural engineer then asked what bearing value he should use in the design and the architect responded with a conservative value which he said that the client had provided. On this basis the structural firm proceeded to design beams, columns and footings for the building.

During the subsequent court proceedings both statements were denied by the client who said that the architect had never asked for a deep foundation study and had never been given a recommended bearing capacity. At the time however the structural engineer had no reason to doubt these statements, particularly since the client had technical expertise on his staff and had experience with other buildings in the area.

Shortly before the contract for construction of the building was let, the client asked the architect for a copy of the soils report. The structural engineer with some indignation told the architect that he should know there was no such report. The architect replied that all the client required was a description of the soils exposed in the test pit and he confirmed this request in a letter to the structural engineer. The employee engineer of the structural firm wrote a letter describing the soils exposed in the test pits and gave an estimate of the bearing capacity. However, his letter did not include a recommendation for deep borings and a proper foundation study.

At the same time that the soils report was requested the client asked the architect to obtain a form letter from the structural engineer certifying that the building was designed according to the requirements of the National Building Code. The structural engineer provided the letter as requested.

The building was built according to plan and shortly after completion showed serious signs of distress. Various alternative explanations were considered and substantial amounts of money were spent in an effort to correct the problem. However, it was clear that the problem stemmed from an inadequate foundation design.

The client sued for damages against the architect, structural engineering firm and the contractor. The litigation was extended and complex. The contractual relationship and responsibilities of the various parties were explored in detail. The case was appealed to the Provincial Court of Appeal and finally to the Supreme Court of Canada.

The contractor was found to have no liability. The architect and the structural engineering firm were found to be jointly and severally liable for the losses suffered by the client. The liability was assigned 60% to the architect and 40% to the structural engineer. However since the architect did not carry any professional liability insurance the entire burden of the losses fell on the structural engineering firm and his insurer.

The architect was found liable because he failed through negligence to fulfill the terms of his contract with the client.

The court concluded that the engineer did not have a contract with the client but he was found liable in tort to the client. The two actions which the court regarded as significant in the decision were:

The soils report which the structural engineer provided failed to estimate settlements in deep soil layers or recommend deep borings.

The form letter stated that the design met all the requirements of the National Building Code. The code requires that for a building of this size the foundation design must be based on a subsurface investigation by a person competent in the field of soil mechanics, or alternatively, be based on local practice including succesful experience with similar buildings and soils in the adjacent area.

Even with the benefit of hindsight there are only a few points at which the structural engineer might have been expected to act differently so as to minimize the risk of things going wrong.

Within the practicalities of the consulting business, it isn't realistic to suggest that he should have refused the assignment. He could only do this if at that time he had such a low opinion of the competence and honesty of the architect that he preferred to forego all future opportunity of paying work with this client.

However, if a letter report on the inspection of the test pits had been written to the architect immediately after the inspection, it would probably have included a recommendation for deep borings and a proper geotechnical study. By the time the letter was written just before the contract was to be let, the lack of a proper geotechnical study was a *fait accompli* and the engineer would feel a great deal of pressure to avoid causing delay and friction by raising this topic which so far as he knew had already been settled by the client.

The structural engineer also assumed more responsibility than he should have in signing the form letter stating that all provisions of the National Building Code had been met. The use of this and similar certificates of supervision and construction are discussed further in Chapter 16 and Loss Control Bulletin No. 66 in Appendix I. In signing this form letter the structural engineer probably felt that he was only referring to the structural aspect of the Building Code while in reality he was taking responsibility for all aspects of the building including those related to foundation design. The structural engineer should have limited his approval only to the portions of the Building Code which applied to his structural analyses.

Erosion slide in a drained natural sand slope after heavy rainfall. Extensive crest and downslope damages caused by leaking storm sewer traversing the slope. If a geotechnical study had been carried out for design and construction of the sewer, could it and should it have contemplated such an event and the extent of the erosion damages?

10 Communications

"When I use a word, "Humpty Dumpty said in rather a scornful tone, "It means just what I choose it to mean — neither more nor less."
Through the Looking Glass

To communicate is to succeed in conveying information. In geotechnical studies the means of conveying information from the consultant to the client includes formal written reports, letters, memos, telephone calls and face to face verbal comments. In any professional liability dispute the written report usually forms a central piece of evidence which may be supplemented or modified by evidence of letters, memos, telephone calls and the recollection of verbal comments.

If the correct information and advice is available, successful communication will involve presenting it to the client in a form which is understandable and unambiguous. This may mean that a different client requires a different style of report. Knowing your client is important. **The non-professional or inexperienced client requires a far more cautionary and explanatory report than an experienced professional engineering organization with whom the geotechnical consultant has been dealing for years.**

The non-professional or inexperienced client requires a far more cautionary and explanatory report than an experienced professional engineering organization with whom the geotechnical consultant has been dealing for years.

Numerous articles, books and courses are available to the geotechnical engineer to assist him in report and letter writing. Report writing is an acquired skill. Some engineers develop it more readily that others but few have it when they first graduate. It is unrealistic to expect a junior inexperienced geotechnical engineer to be a skillful report writer. Establishing a procedure for review of draft and final reports is an important defence against professional liability claims. In addition to clear unambiguous writing, however, there are some aspects of a geotechnical report which may be important in defending against a professional liability claim.

Proposal

If the study was based on a proposal it should be identified in the report and if appropriate, it may be included as an appendix. If the study that was actually done differs from what was anticipated in the proposal, the changes and the reasons for the changes should be identified. If there is no written proposal the understanding on which the study was based should be described.

Authorization

The authority on which the study was undertaken should be identified as well as the authorization for any change in the scope of the study. The person to whom the report is addressed should be identified.

Project

The project for which the study was undertaken should be described in reasonable detail with reference to any drawings or reports. If the project has changed during the course of the study the change should be described.

Purpose

The purpose of the study should be described and purposes for which the study is not appropriate should be mentioned.

Scope of the Work

The work actually done should be described including test borings, laboratory tests, equipment used, etc. Some of the descriptions may consist of plans and sketches.

Presentation of Data

Borehole logs showing sample depth, sample type, and laboratory identification tests should be included to the extent necessary to justify the conclusions and recommendations and to provide a permanent record of the work done. There is some risk, from professional liability point of view, of providing too much factual descriptive data. Resist the temptation to pad up the report with extraneous data. In a dispute the opposing side may "nit-pick" one or more points in an attempt to discredit the accuracy of the data.

As a general rule it is desirable to present technical data in the report in only one form. For example a soil profile may be presented on a detailed borehole log. If it is also described in detail in the text and there are minor discrepancies between the two, these discrepancies may be blown out of all proportion to confuse the issue in a dispute. In any case the plaintiff will use the description which is most damaging to the case of the geotechnical engineer and the plaintiff's lawyer will use the difference to create confusion in cross examination.

Jargon

Many of the words used by engineers have very special and limited meanings which are different from the dictionary definitions or the meanings commonly understood by laymen. If these words are used in the report, particularly in the conclusions and recommendations, they may lead to an unfavourable interpretation by the courts in a professional liability case.

Engineers tend to use absolute and positive statements in every day communications. Such phrases as, "These results undoubtedly show" or, "if this is done no further problems will develop" are not uncommon. They are the result of a natural desire to convince and reassure the client. However such phrases when used in reports may give an unrealistic indication of certainty and precision. Such certainty and precision is seldom justified in geotechnical studies. Flamboyant or complex sentences and superlative and absolute adjectives may easily lead to unjustified expectations on the part of the client and these expectations may lead to professional liability claims and undesireable decisions in court. They have no place in a design professional's objective report.

Many of the words used by engineers have very special and limited meanings which are different from the dictionary definitions or the meanings commonly understood by laymen.

Reaching the Right Party

The preceding comments have dealt with aspects of writing geotechnical reports. From a professional liability point of view, there are several administrative aspects of communication which must be kept in mind.

A report or letter may contain all the essential information and advice that the geotechnical engineer is responsible for providing. It must, however, be delivered to the person or organization to whom the geotechnical engineer is legally responsible. If it is addressed to the wrong organization and a professional liability claim arises in which the geotechnical engineer's defence is that the client did not follow his advice he may be found liable on the technicality that he cannot prove that his client ever received his advice.

There are two common situations where some variation of this administrative problem can develop.

The first of these is in a large and complex project where there are a number of players including perhaps an architect, a structural engineer, a project manager, a site supervisor, an owner, a general contractor and several subcontractors all of whom have some interest in the advice and activities of the geotechnical engineer. This is a perilous situation for the geotechnical engineer. As soon as the character of the project becomes clear the geotechnical engineer should establish who will be responsible for paying his bills and get that organization to name an individual as the point of contact who will be responsible for requesting and authorizing any geotechnical studies or reports required for the project. This same individual should then be responsible for receiving the advice of the geotechnical engineer and distributing it within the project team.

Needless to say being employed by two different members of the project team is a perilous situation. During construction of some dykes a geotechnical firm found that one office was providing a technician to inspect fill placement during the day paid for by and reporting to the project manager. Another office of the same firm was providing a technician to inspect fill placement during the night shift paid for by and reporting to the contractor's superintendent. There was a personality conflict between the project manager and the contractor's superintendent so they were not communicating with each other. Luckily the technicians took it upon themselves to keep in close touch and provide a link between day and night shifts which was missing in the project administration. Needless to say a hazardous situation for the geotechnical firm.

A second situation where confusion as to who is legally the client often arises in dealing with a builder/developer. Many of these organizations maintain separate companies for each project often with separate companies providing design, construction and management. The initial request for a geotechnical study may come from the developer company at the stage of feasibility or property acquisition and later requests for advice may come from the designer or builder companies. Often the names of the companies are similar as for example Sunshine Builders Ltd., Sunshine Developers Ltd. and Sunshine Managers Inc. Frequently the same individuals wear different hats throughout the organization. Make sure who you are in a contractual relation with and what services he expects to receive from you.

A comprehensive report of substandard density tests is a liability if it is delivered when the building is half completed.

This aspect of communication is very difficult for the geotechnical engineer. Aside from professional liability considerations however, it is worthwhile for the geotechnical engineer to determine who is his client. A few unscrupulous developers use a technique of shuffling a deck of companies so as to conveniently leave the geotechnical engineer's bills in a company with no assets.

Not all geotechnical advice is transmitted by reports and letters. Frequently a geotechnical engineer will be asked for comments, information or advice in a telephone call. During site visits the geotechnical engineer will make verbal comments on his observations. At the very least these telephone and verbal comments should be confirmed in a memo to the engineer's file at the time they are made. A preferred procedure is to confirm these comments with a written memo to the client with a duplicate retained in the engineer's file.

Letters and memos to confirm advice and observations should be written and delivered immediately. Not only do memories fade but circumstances change and advice must be acted on within the context in which it is made. **A comprehensive report of substandard density tests is a liability if it is delivered when the building is half completed.**

11 Case History V

This case history illustrates the hazard of filling a report with an excess of detail and comments.

A major high-rise office building with several levels of underground parking was planned for an urban development. A geotechnical firm was employed to carry out and report on subsurface conditions. Test drilling established bedrock (a horizontally bedded sedimentary rock) at a shallow depth and the borings were extended to the full depth of the proposed excavation. A professor of geology was retained by the geotechnical consultant and asked to examine and describe the core. His report was very thorough and comprehensive. The age, lithology, structural discontinuities, mineralogy, jointing, bedding, and fossils were described and discussed in detail, even though the report was based on an examination of discontinuous small diameter core. The entire geological description was incorporated in the geotechnical report which became part of the contract documents.

In the course of drilling and blasting the bedrock to excavate for the basement and footings, considerable overbreak occurred which the contractor was obliged to backfill with lean concrete. Blasting was carefully controlled by an explosives expert hence the overbreak could not be attributed to poor procedures.

The contractor claimed for an extra as a result of the overbreak and his "expert" claimed that the contractor relied on the geological description of the core as "thinly bedded" and accordingly made little allowance for overbreak. Photographs taken during construction showed horizontal beds 1.0 to 1.5 meters thick which in terms of mass rock would not be regarded as thinly bedded.

It was concluded that the contractor had a valid claim and was paid an extra. This claim might not have been allowed if the report had merely reported the rock type, elevation and percentage of core recovery (RQD) and included representative photographs of the core.

From a practical point of view the small diameter of the core made it unsuitable for determining the spacing of the bedding planes. It is unlikely that the geologist or the geotechnical engineer anticipated that the rock description would be used to predict the behaviour of the rock when excavated. If the professor had realized the importance that would be attached to the term "thinly bedded" he would no doubt have considered the use of the term more carefully. An examination of a nearby rock outcrop would probably have been more informative than the core fragments. A cynic might suspect that the contractor only studied the description of the rock in detail when he realized the extent and cost of the overbreak.

When you pad out a report with a mass of extraneous detail and comments, you are providing answers to questions which have not been asked and the answers you have given may well be wrong or misleading.

12 Limitation of Liability and Disclaimer Clauses

Acheson's Rule of the Bureaucracy — a memorandum is written not to inform the reader but to protect the writer. quoted in The Official Rules.

"The dealer and processor shall not be liable for any loss or damage to film or pictures while in the custody of them or their agents because of negligence or any cause except that any film or pictures so lost or damaged will be replaced with similar unexposed film."

This, or a similar statement appearing on a package of film limits the liability of the laboratory which processes the film. It is generally accepted by the purchaser, because he has no alternative and because the loss of an exposed roll of film can seldom be shown to involve a significant monetary loss.

In a competitive marketplace purchasers of a foundation investigation for multimillion dollar projects are less likely to accept a limitation of liability clause in their contract.

Nevertheless both the Association of Soil and Foundation Engineers in the United States and the National Program Administrator Inc. in Canada recommend that a consulting engineer endeavour to limit his liability in his contract. This limitation might be achieved by the insertion of the following clauses in the contract.

"The client agrees that any claim which it has or hereafter may have against the Geotechnical Engineer in respect of the Services, howsoever arising, whether in contract or in tort, shall be limited to the amount of the Geotechnical Engineer's professional liability insurance in effect at the date of the execution of this Agreement, including the deductible portion therein."

"The Geotechnical Engineer's professional liability insurance policy is available for inspection by the Client at all times upon request. Prior to the date of the execution of the Agreement, if the Client wishes, because of the Client's particular circumstances or otherwise, to increase the amount of the coverage of such policy, or to obtain other special insurance coverage, then the Geotechnical Engineer shall co-operate with the Client to obtain such increased or special insurance coverage, with the cost of such increased or special coverage, to be a Reimbursable Expense."

Such a limiting clause is probably never included in a letter proposal/contract which is the most common agreement with a client for geotechnical services. Discussion of such a clause with the client would at least bring into focus the magnitude of the responsibility of the geotechnical engineer when compared with the dollar value of his fees.

Although the geotechnical engineer may have some difficulty in negotiating a limitation of his liability in contract, he can set up a a line of defence by judicious use of various clauses in his report on the study in question. These clauses may be grouped under the general category of disclaimers. They are statements which clearly define limitations in the report which are frequently not understood by the client, and in some instances not anticipated by the consultant.

The first line of defence is a geotechnical report which is carefully written, carefully reviewed, carefully edited and carefully proof-read. The requirements for a good geotechnical report are described in more detail in chapter 10 on Communications.

Geotechnical engineering is carried out in an environment which differs significantly from the environment of other engineering and architectural disciplines and differs radically from the world as perceived by a layman. The framework of this environment is recognized and taken into consideration by a competent experienced geotechnical engineer and should be defined for the client as part of the geotechnical report. This includes the knowledge that:

- Subsurface conditions vary over a site.

- Test borings indicate approximate conditions only at the borehole locations.

- Certain soil deposits, such as glacial tills, likely contain cobbles and boulders, even if not directly encountered in a test boring.

- Accurate definition of bedrock surface, or contact between shattered and sound bedrock, is sometimes not possible using routine test boring procedures.

- The amount of detailed information obtained at a bore hole varies with the equipment and procedure.

- Subsurface information gathered for one purpose (e.g. pile design) is not suitable for another (e.g. design of a retaining wall).

- More subsurface information or better quality information costs more money.

The soil and groundwater conditions observed during the subsurface investigation may change with time.

The character and properties of the soil or rock may change when exposed to the elements of the weather, drying, wetting and freezing.

The character and properties of the soil or rock may be altered by construction activities, including excavation, traffic, compaction, pile driving and blasting.

At the stage of preparing his report the geotechnical consultant may not always be aware of final planning with regard to construction procedures or permanent pumping installations, which may cause adverse effects on either the contemplated works, or adjacent property. If he is experienced he will recognize potential risks and the limitations of his work and provide such general comments in his report, with a recommendation for further geotechnical follow-up as planning becomes more advanced. Where appropriate he should also recommend additional fieldwork or installation of monitoring instrumentation to quantify certain aspects, if beyond the scope and budget terms of his initial mandate.

The description of the limitations of the report can be incorporated as statements or comments scattered throughout the report or included as a one or two page summary of conditions, bound with the report and referred to in the text.

The following is a suggested format which could be incorporated in whole or in part in both the proposal/contract and the geotechnical report. The wording should be reviewed and modified as required by the geotechnical firm's legal counsel.

Geotechnical Report

General Conditions and Limitations

Use of the Report

The factual data, interpretations and recommendations contained in this report pertain to a specific project as decribed in the report and are not applicable to any other project or site. If the project is modified in any significant way or if the project is not initiated within eighteen months of the date of the report (name of firm) should be given an opportunity to confirm that the recommendations are still valid.

Description of Soil and Rock

Soils and rock descriptions in this report are based on commonly accepted methods of classification and identification employed in professional geotechnical practice. Classification and identification of soil and rock involves judgement and (name of firm) does not guarantee descriptions as exact but infers accuracy only to the extent that is common in current geotechnical practice.

Logs of Borings and Subsurface Interpretations

Soil and rock formations are variable to a greater or lesser extent. The boring logs indicate the approximate subsurface conditions only at the locations of the boreholes. The precision with which subsurface conditions are indicated depends on the method of boring, the frequency of sampling, the method of sampling and the uniformity of subsurface conditions. The spacing of boreholes, frequency of sampling and type of boring have been selected to meet the needs of the project within constraints of budget and schedule.

Subsurface conditions between boreholes are inferred and may vary significantly from conditions encountered at the borings.

Groundwater Conditions

Groundwater conditions described in this report refer only to those observed at the place and time of observation noted in the report. These conditions may vary seasonally or as a consequence of construction activities on the site or adjacent sites.

Soil and Rock Conditions

The soils and rock conditions described in this report are those observed at the time of the study. Unless otherwise noted, those conditions form the basis of the recommendations in the report. The condition of the soil and rock may be significantly altered by construction activities (traffic, excavation, pile driving, blasting, etc.) on the site or on adjacent sites. Excavation may expose the soils to changes due to wetting, drying or frost. Unless otherwise indicated the soil must be protected from these changes during construction.

Changed Conditions

Where conditions encountered at the site differ significantly from those anticipated in this report, either due to natural variability of subsurface conditions or construction activities, it is a condition of this report that (name of firm) be notified of the changes and provided with an opportunity to review the recommendations of this report. Recognition of changed soil and rock conditions requires experience and it is recommended that an experienced geotechnical engineer be employed to visit the site with sufficient frequency to detect if conditions have changed significantly.

Drainage Systems

Drainage of subsurface water is commonly required either as temporary or permanent installations for the project. Improper design or construction can have serious consequences. (name of firm) can take no responsibility for the effects of drainage unless we are specifically involved in the detailed design and construction monitoring of the system.

Standard of Care

Services performed by (name of firm) for this report are conducted in a manner consistent with that level of skill and care ordinarily exercised by members of the profession currently practicing under similar conditions.

Mephisto's UNETHICAL ADVICE

A regular feature of the Devil's Advocate and the Hades Daily News

Dear Mephisto:
I'm a contractor's superintendent who is being hassled by the city's building inspector because he says I have undercut the slope next to the street. The geotechnical firm's report recommended excavating it at 1/2 horizontal to 1 vertical or installing shoring. Everyone knows you can't afford to do that.

Anyway, last week there was a young engineer from the geotechnical firm on site and I said "I'll give you a hundred dollars if you'll write a letter saying that the slope is ok." He looked a bit startled and the next thing I know the building inspector insisted I haul in gravel to backfill the slope and get a structural engineer to design shoring.

Where did I go wrong? Maybe I should have offered him $500.

Yours truly

Going broke in a hurry

Dear Going Broke:

These heavy handed tactics never work with an idealistic young engineer. These lads don't know much but they do know a bribe when they see one.

A technique I have used successfully from time to time is to phone up the young engineer and say you're a bit uneasy about this slope and will he come out and look at it. Don't mention the building inspector. He'll probably recognize that it is steeper than was recommended but after all 90% of those slopes will stand vertical for a few days to a week or so. He will be flattered at your asking him and after all being helpful is the essence of consulting. So he will probably say "Well it is pretty steep but at this stage the best thing to do is to get the wall up and braced and backfilled as quickly as possible". Then when the building inspector calls you can say you are following the engineers advice and a couple of weeks later when the bank falls in with the gas and sewer lines you can claim it was the engineer's recommendation and his insurance should pick up the tab.

Better luck next time.

Mephisto

P.S. Be sure to ask for the junior engineer, seniors may have been through this exercise already.

13 Case History VI

This case history illustrates the problem caused by a simplistic interpretation of subsurface data. It is also an example of a common failure to make use of all available information.

The client, a municipal council, employed a consultant in municipal engineering to design and supervise the construction of a wastewater treatment plant and sewage collection system. The municipal consultant employed a geotechnical sub-consultant, and sent the geotechnical firm a drawing showing a profile of the main sewer line with instructions to put down boreholes or test pits at intervals of 150 meters. The investigation was to extend 0.5 meters below the grade line of the sewer.

The alignment of the sewer was staked in the field and the geotechnical firm put a technician in charge of the work to arrange for the drilling and test pits as required, to collect samples and to compile the results.

Over most of the length of the sewer, granular soil was encountered to the full depth of the test pits and borings. In one section two test holes encountered rock 1.0 meters above the proposed grade.

The geotechnical consultant submitted a copy of the profile showing stratigraphy at the test locations. The boundary between bedrock and granular overburden was drawn as a straight line between the two test holes. The drawing was approved and signed by the geotechnical engineer.

The municipal consultant prepared tender drawings and specifications. The tender form included a table of estimated excavation quantities of soil and rock which the municipal consultant had calculated from the information submitted by the geotechnical consultant. The contract was a lump sum for the quantities given in the tender, and unit prices were requested for any additional quantities over and above those included in the lump sum.

Several contractors bid on the work but the low bidder was a local contractor who knew the area well. His tender included a very high unit price for extra rock excavation. In the tender quantities, only a small amount of rock excavation was indicated. The contractor had walked the line in the field and recognized areas of rock outcrop 20 to 30 meters outside the alignment of the sewer and he suspected that the tender documents substantially underestimated the quantity of rock to be excavated.

During the course of construction a knob of bedrock was encountered between the two boreholes where rock was shown only slightly above the bottom of the excavation.

The unanticipated rock combined with the high bid price for extra rock excavation resulted in a cost which exceeded the total of the original lump sum bid. The municipality was very upset at this cost overrun and claimed that the prime consultant and the geotechnical consultant had made an error and were liable for the extra costs.

The geotechnical consultant claimed that the information was correct at the borehole and test pit sites and that since the location and spacing of the program had been imposed on him together with a very tight budget, he should not be held responsible for the use made of his work.

The prime consultant claimed that he was entitled to rely on the interpretation of subsurface conditions provided by the geotechnical firm since the geotechnical consultant was an expert in these matters and it was the geotechnical consultant who had connected the borehole data with straight lines.

Fortunately the local contractor was anxious to maintain good relations with the municipality and its consultants and agreed to re-negotiate the price for extra rock excavation to a more reasonable level. Both the prime consultant and the geotechnical consultant suffered a loss of credibility with the municipality.

The geotechnical engineer by connecting the borehole data with straight lines gave an unqualified interpretation of subsurface conditions when he wasn't asked to interpret subsurface conditions at all. The prime consultant might have got a better result if he had asked the geotechnical firm to plan the subsurface investigation and provide an estimate of quantities for excavation though this would have undoubtedly involved more fees for the geotechnical firm.

This case history also illustrates the tendency of many engineers to rush directly into a drilling program without bothering to gather the abundant information which is lying around on the surface. Most clients are willing to pay to mobilize drill rigs and other machinery but are less willing to pay for an intelligent observer to look at a few airphotos and spend a few hours walking around a site. A similar situation exists in studies of building sites where a modest expenditure to gather the history of the site and previous developments is invaluable in improving the planning and interpretation of a drilling and testing program.

14
Hazardous Clients and High Risk Situations

"Who are those guys?" — Butch Cassidy and the Sundance Kid.

At times the geotechnical engineer may feel like Butch Cassidy, he is pursued by an implacable posse of faceless riders, consisting of clients and their lawyers.

These clients, however, are not completely unknown. Every article on loss prevention in the field of professional liability identifies groups of clients whose projects present a higher than average risk of professional liability claims.

Unfortunately, in the real world, few geotechnical firms can afford the luxury of rejecting assignments from these high risk clients. Furthermore, many of these clients are in very great need of the services of a competent geotechnical engineer.

As is the case with many problems, recognition of the problem is half way to the solution. Only in a few instances is refusing to work for the client the only solution. The adoption of a variety of defensive measures, most of which are good business and management practices, is usually sufficient to reduce the risk of a professional liability claim to an acceptable level. Of course, such a risk can never be completely eliminated.

Developers of Residential and Commercial Properties

At the top of everyone's list of high risk clients are developers of residential and commercial properties, particularly if they are underfinanced. There are a number of reasons why these clients are high risk.

To succeed developers must be optimists and as optimists they tend to

discount the geotechnical engineer's presentation of risk and uncertainty. Furthermore the developer seldom has an appreciation of the conditions and procedures involved in geotechnical engineering. If he has been involved in a number of projects he may have developed a fair amount of experience in discussions with architects and structural engineers. But he will expect all sites to be generally similar and equivalent to the better sites in his recent memory. He does not wish to hear about poor soil conditions and site preparation requirements that will involve greater costs than he has allowed in his budget.

The geotechnical engineer presents a proposal for a preliminary site evaluation at a cost of $1000 and indicates that the results may dictate further investigations and field supervision to a total of $5000. The developer only hears and registers $1000.

To succeed developers must be optimists and as optimists they tend to discount the geotechnical engineer's presentation of risk and uncertainty.

The financing procedures of the developer also exert great pressure on him to restrict the extent of the geotechnical investigation. Generally the geotechnical investigation is part of the feasibility study and as such is part of the front end loading of the project before mortgage financing is obtained. Therefore, the cost of the geotechnical investigation comes directly out of the pockets of the developer and his partners.

The developer, because of his lack of appreciation of geotechnical engineering, is frequently willing to accept risks without properly appreciating their significance. Finally, many developers accept risks anticipating that they will be passed on to the ultimate owners. In these circumstances the ultimate owners may endeavour to recover the financial consequences of these risks by claiming against the developer's consultant.

Volunteer Organizations

A second group of high risk clients can be lumped together under the general heading of volunteer organizations. These include the councils of small municipalities, boards set up for water and sewage districts, groups endeavouring to establish recreational facilities such as curling rinks, swimming pools, etc. These clients have three characteristics which contribute to the high risk character of their projects. In general they operate on a limited budget which is closely scrutinized by their electorate. On the average they are unsophisticated so far as the cost and value of engineering services is concerned. They know the difference between $100 and $1000 but not between a hundred dollar job and a thousand dollar job. Finally, the membership of these boards and councils changes from year to year. The geotechnical engineer will deal with one group when the project is being set up and with a completely different group when it is finally completed.

A special group of high risk clients for the geotechnical engineer are those clients who are involved in and subject to political pressures. This situation arises where the geotechnical engineer is subconsultant to a general consultant who is involved in promoting a particular project with the owner who is frequently, though not always, a small municipality. Within the owner's camp there is a group opposed to the project. Within the group in favour of the project, which includes the prime consultant, there is strong pressure to minimize or conceal any observations or conclusions which are detrimental to the project. The geotechnical engineer is caught in a political situation where his concerns and recommended courses of action cannot be evaluated on a rational basis.

Project Management

In recent years a number of techniques for efficient and accelerated handling of projects have been developed. These include fast tracking, design/build, and project management. When properly handled these procedures provide significant benefits to the owner by expediting completion of the project and significantly reducing financing costs. Without exception they increase the risk to the geotechnical engineer, particularly where he reports to the new breed of project manager, whose skills and training are in business administration and the production of computerized job data outputs, but not necessarily in a good understanding of site preparation and construction methods. His primary interest is to control costs and schedule.

All of these methods of handling projects involve the expectation that decisions regarding design and construction will be made as the project proceeds. The number and significance of these decisions is of a different order of magnitude when compared with the changes involved in administration of a normal contract.

Although nominally part of the team set up to handle fast tracking, design/build, and project management procedures, the geotechnical engineer seldom continues to participate in the team's deliberations throughout the project. His involvement usually terminates at an early stage after he has produced a report on site investigations. This is not surprising since only a few of the decisions at the later stages have any direct bearing on geotechnical considerations. However, the result will be that decisions will be made by others based on their interpretation of the geotechnical report or, more likely, without any appreciation of the significance of the geotechnical data relative to the proposed change. Where the importance of geotechnical input is recognized the geotechnical engineer will often be called and asked to express an opinion without being given an opportunity to familiarize himself with the current status of the project.

Under these circumstances the geotechnical engineer must make a special effort to avoid being held responsible for decisions made by others based on an interpretation of the wording of his earlier reports. When he is called in to give an opinion on some proposed change he must ensure that he fully understands the current status of the project and the possible ramifications of his recommendations through the project.

Difficult Sites and Unusual Projects

Difficult sites and unusual projects create high risk clients. Difficult sites present a challenge to a geotechnical engineer and he may enthusiastically embark on a study which he regards as an opportunity to apply advanced technology and exercise his technical skills to a high degree. For the owner however a difficult site represents only unexpected costs, problems and delays which he may blame on the geotechnical consultant. When he is faced with a difficult site or an unusal project therefore, the geotechnical engineer must apply a high level of technical skill but even more important he must offer a very high level of communication and administrative skills.

Analysis of the stability of a natural or artificial slope which has not yet failed is an example of a difficult geotechnical problem. The analytical procedures require simplifying assumptions regarding stratigraphy, soil strengths and groundwater conditions to produce numerical values of safety factor. The numerical values are easy to communicate but the uncertainties are easily lost and the engineer's client may feel assured that the slope is, without reservation, safe. In this situation the engineer must have great skill in communicating the meaning of the results and portraying the risks in true perspective. As a check **the engineer should ask himself two questions: "What are the consequences if I am wrong?" and "How wrong can I be?"**

The engineer should ask himself two questions: "What are the consequences if I am wrong?" and "How wrong can I be?"

Remote sites including foreign assignments should always be regarded as having potentially high risk. The site investigation is more difficult. If the drill rig or sampling equipment is inappropriate it is difficult or impossible to change. The crews are anxious to get home. Supervision is difficult. There is no background of local experience. All of this adds up to a greater than normal chance of deficiencies in the field data. Furthermore, the consequences of an error are likely to be much more costly to correct in a remote site. Accordingly, the engineer must try to compensate for this high risk situation in every way he can.

Litigious Clients

A few clients have a reputation of being prone to resolve any difficulty (including paying their bill) with a threat of a lawsuit. As soon as they are recognized, assignments from these clients should automatically be refused. Probably the most dangerous client you will encounter is an underfinanced developer whose brother-in-law is a lawyer.

Others

Finally a very hazardous client is one who doesn't understand anything about geotechnical engineering and treats the professional consultant as a contractor or material supplier. In requesting a foundation investigation he will say "Just give me a bearing capacity number that I can use" in about the same way that he would order a load of cement from a supplier. Unfortunately a certain number of architects and engineers fall into this group and if you are unable to persuade them of some of the pitfalls of this approach you are better to avoid their assignments. Otherwise you may find yourself responsible for their misuse of your reports.

Roger Goldie in his book "Muddling Through the Art of Properly Unbusinesslike Management" has a chapter entitled Dealing with Experts, the Art of Managing from ignorance. This deals with the problems faced by a manager (the client) who must use the service of an expert (the geotechnical engineer).

From the point of view of the client there are a number of things that he must do in order to get a satisfactory result from the geotechnical consultant. The way the client deals with the consultant contributes to the end results as much as the expert's professional knowledge or skill. The client must be competent in his own specialty and must exercise proper management skills in his dealing with the geotechnical consultant.

The following are some of the ways a client fails in his dealings with the consultant.

- puts off employing a consultant so that there is no time to do a proper job.

- fails to inform the consultant of the background of the project which may affect the use of the consultant's advice.

- fails to establish schedules and budgets.

- has unrealistic expectations of what the geotechnical consultant can do.

- endeavours to second guess the consultant by specifying the number of boreholes or otherwise arbitrarily limiting the scope of the investigation.

- endeavours to oversimplify the problem or the solution.

It is impossible to do a good job for an incompetent client.

If the geotechncial consultant recognizes these deficiencies in the client he can usually compensate for them though it will make the study more expensive and the solution more conservative.

In general, **it is impossible to do a good job for an incompetent client.** This rule probably applies equally well to other fields of professional consulting including medicine, law and accounting.

To be successful, the business of consulting must be carried out in an atmosphere of mutual trust and respect. At times it is difficult to distingush between incompetence and dishonesty.

For a geotechnical consultant, utopia would be a world where he was able to decline assignments from any client whom he recognized as hazardous. In the real world of carrying on a successful and profitable business the geotechnical consultant seldom enjoys this luxury. The recognition of a hazardous client serves as a warning signal to proceed with caution. The litigious client can be rejected outright. The underfinanced client can usually be rejected since the consultant will have difficulty collecting his fee even if he doesn't suffer from a professional liability claim. Other hazardous clients may be acceptable, but the consultant must take special care in drawing up his contract, writing his report and supervising his staff in order to minimize the risk of a professional liability claim.

Result of improper techniques during placement of moisture sensitive clay fill.

Did the geotechnical consultant adequately emphasize the need for detailed specifications on the selection and handling of the material at the borrow source, lift placement and compaction procedures, run-off drainage control and restrictions on wheeled equipment trafficability to achieve the desired end result for the compacted fill?

15
Case History VII

This case history illustrates the situation which can develop when a difficult site meets a project which cannot afford to resolve the problems presented by the site conditions.

A low lying site on the outskirts of a major city was chosen for a large shopping centre development. The geotechnical consultant retained by the developer found the subsoils at the site to consist of 20 meters of soft sensitive clay overlying dense glacial till. The consultant recognized the poor foundation conditions and recommended that the structure be founded on end-bearing piles driven to refusal in the glacial till. Although this increased the costs over a foundation on spread footings, it was considered acceptable for the columns and walls. However, the project would have been uneconomic if the floor were designed as a structural slab supported on piles.

Accordingly the consultants were under considerable pressure to come up with a grade-supported slab on fill. The consolidation tests on the clay indicated some minor pre-consolidation about equal to the loading of the proposed fill and floor. It was decided to go ahead with the non-structural slab on grade.

After the structure was completed and the stores occupied, the effects of settlement began to appear. The floors sagged significantly between column pile caps, sliding glass doors cracked or could not be closed, freezer counters had to be shimmed and cracks appeared in floor-supported partitions. After 18 months differential settlements amounted to 75 mm.

At this point the geotechnical consultant was viewed as the culprit and an outside expert was retained to examine the cause of the problem and advise regarding a claim for negligence.

In hindsight, the geotechnical consultant had been pressured by the economics of the project into gambling that his laboratory tests and analyses accurately predicted the behaviour of the soft clay under full scale field conditions. The amount of testing and analysis was limited. The developer naturally pressured the geotechnical engineer to adopt an optimistic view in order to keep the project alive. When time failed to bear out the optimism, the developer was obliged by the economics of the situation to try to recoup his losses by whatever means possible.

This is a typical situation where the geotechnical engineer should ask himself "How wrong can I be?" and "What are the consequences of being wrong?"

It is quite possible that at the time the decision was made to use a slab on grade, the geotechnical engineer in his best judgement, and based on past experience, believed that there was a good chance that the slab would perform satisfactorily. However, when the outside expert is brought in he is 100 percent sure that the slab is not performing satisfactorily. With the benefit of this knowledge and a much larger budget for testing and analysis he will have no difficulty in identifying all the reasons why it was a wrong decision to use the slab on grade. These are the reasons which will be used to support a claim against the geotechnical engineer.

16
Inspection, Supervision and Certification

Nature always sides with the hidden flaws. Hidden flaws are always found only at the time you can least afford the discovery. - quoted in The Official Rules.

A significant proportion of professional liability claims fall into the general category of "provision of field services". Field services provided by geotechnical firms include such things as compaction control, inspection and evaluation of soils exposed during construction and inspection of pile installation. These services would be described in a contract under the general headings of supervision, inspection and certification.

There are a number of reasons why field services create professional liability problems for geotechnical engineers as well as for all engineering consultants.

- Usually insufficient funds are allotted to pay the geotechnical engineer to do a proper and thorough job.

- Many geotechnical firms regard these services as routine "bread and butter" jobs which are assigned to technicians and junior engineers without adequate supervision by experienced senior personnel.

- In many cases the cost of these services is included as a lump sum payment in the construction contract and the contractor then employs and directs the geotechnical inspector as a subcontractor. This places the geotechnical firm in an intolerable conflict of interest situation.

- Frequently the need for field services is intermittent and the responsibility for requesting site inspection is given to the contractor or to a site representative of the owner. As a result the geotechnical engineer may not be on site at a critical stage of the construction.

- Many municipalities request various forms of certification to be completed by a professional engineer at various stages of the approval process. These certificates place the responsibility for the contractor's workmanship on the professional engineer to a degree that his assignment from the owner would not permit him to accept.

As with so many other professional liability problems, in most cases the problems related to field services start with the original assignment to the geotechnical engineer.

The assignment for the foundation investigation is carefully drawn and a carefully written report is submitted. Then several months later the engineer gets a phone call. "The contractor is placing some fill at the site would you send someone out to look at it". This then develops into intermittent site inspection and at the end of the project the engineer is presented with a request to sign a certificate assuring the owner that compaction and materials are in accordance with the design and specifications.

There have been extensive discussions about the meaning of inspection and supervision and whether continuous or intermittent attendance at the site is intended. In general the courts do not differentiate between supervision and inspection and do not consider the amount paid for the services as indicating the level of services which should have been provided.

In order to minimize the hazard of professional liability claims arising from field services, the geotechnical engineer should ensure that there is a clearly understood agreement between himself and the owner indicating what he will do, what it will cost, what are the uncertainties and risks and what limitations there will be on any certification. If the project is to be passed on to third parties, as in the case of a housing development or municipal services, the engineer must be doubly cautious to ensure that he is not going to be responsible for construction deficiencies over which he has no control.

The engineer should keep in mind that he cannot certify what he has not seen and that if he does he may be found guilty at least of professional negligence or possibly of fraud.

17

Case History VIII

Letters from a Mother's Scrapbook

May 30

Dear Mom,

Well here I am in the big city ready to go to work. I won't be sorry to be finished with professors, lab reports and exams and start earning some money to pay you back that loan for the last year at university.

I'm going to enjoy using my new SLR camera and will send you some photos of my first project.

Your loving son

Freddy

BASc.MSc.(Soils)

Pretty impressive eh.

June 10

Dear Mom,

Well I've been at work for a week now and I think I'm going to enjoy working for Crossgrain and Associates. This is a branch office so we only have a few engineers and technicians and the laboratory is pretty minimal compared to what we had at university, however we may be getting personal computers before too long.

Mr. Hardwood is the partner in charge of this office and he seems very nice though he is very busy.

Your loving son

Freddy

June 15

Dear Mom,

Well I've got my first assignment but so far haven't been out in the field so I can't send you the picture I promised. Mr. Hardwood gave me the file to review and I'm to go in to talk to him tomorrow.

Last fall the company did a study of the site which is about 32 kilometers (20 miles for you mom) west of here. This company is going to build a warehouse to store their supplies, paint I think, when it comes in by train from their factory. Anyway we drilled four holes and found about twelve feet of soft peat about the consistency of manure but doesn't smell so bad. We could drive piles but have decided to go for a preload fill. A preload is piling up dirt so that the weight squeezes water out of the peat. When the preload has settled enough you remove more than the weight of the building and then when you build the building it doesn't settle anymore. We took this in professor Chiang's class and I'm going to enjoy seeing how it really works.

Your loving son

Freddy

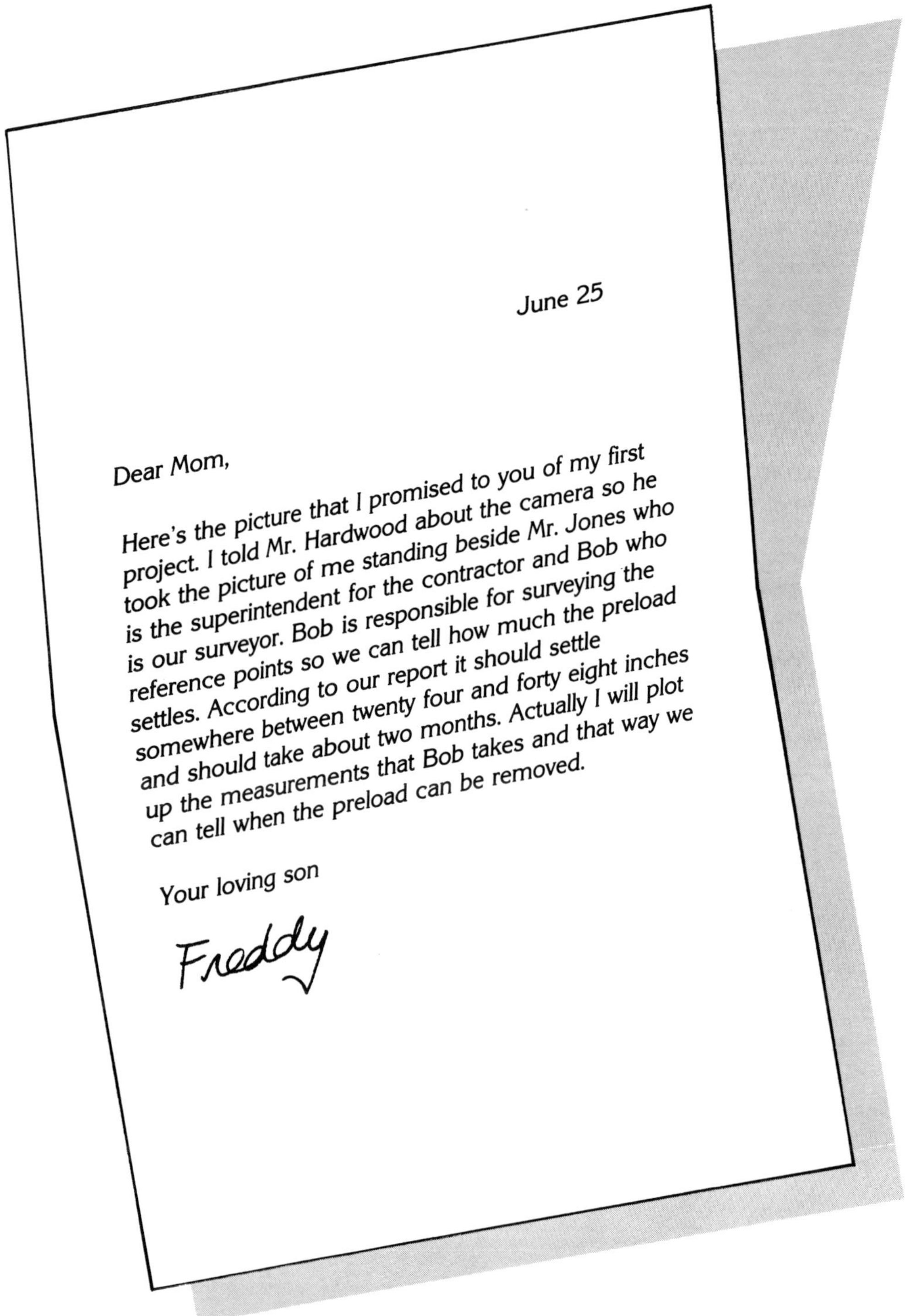

June 25

Dear Mom,

Here's the picture that I promised to you of my first project. I told Mr. Hardwood about the camera so he took the picture of me standing beside Mr. Jones who is the superintendent for the contractor and Bob who is our surveyor. Bob is responsible for surveying the reference points so we can tell how much the preload settles. According to our report it should settle somewhere between twenty four and forty eight inches and should take about two months. Actually I will plot up the measurements that Bob takes and that way we can tell when the preload can be removed.

Your loving son

Freddy

July 5

Dear Mom,

It was too bad that it rained all last week. It sort of spoiled the celebrations. It also interfered with placing the preload. Because of the rain it was difficult for the contractor to compact the fill. When Mr. Hardwood went out he made them dig out the soft spots. They weren't very happy about it. However things are going better now.

Your loving son

Freddy

July 12

Dear Mom,

Boy has it been raining. I thought it would never stop so I went and bought myself one of these yellow slickers - pretty colourful for me. As a result the roads have been so muddy that the trucks haven't been able to haul in the preload fill let alone compact it.

Give my best regards to the ducks - they should be happy.

Your loving son

Freddy

July 20

Dear Mom,

Well the weather finally cleared up but if its not one thing its another. Mr. Chequers is the new manager for the paint company and he turned up the other day mad as a wet hen about the bill. He's a young fellow not much older than me and just got his MBA (that's master of business administration) last year. He stomped up and down and swore quite a bit and said he wasn't going to pay any $75 an hour for Mr. Hardwood to come out on his site. I said he'd have to speak to Mr. Hardwood about it because I didn't have anything to do with the billing. It was pretty unpleasant. Anyway Mr. Hardwood called me in and said he probably wouldn't spend as much time on the site now that things were going and for me to keep an eye on things and call him any time I thought I should. Anyway the warm weather makes me feel better and the boys at the shop are planning a barbecue this weekend.

Your loving son

Freddy

August 5

Dear Mom,

Well the fine weather has held and we got the preload fill completed by the last day of July. Bob put out his settlement hubs and surveyed them in. When I plotted them up the fill didn't seem to be quite high enough but Mr. Jones who is the contractor's super said it was probably because settlement had occurred while the fill was being placed. In any case they had hauled in the amount of fill they had estimated for the preload. Mr. Hardwood is away this week at a meeting in Ottawa. I must remember to check with him next week to see if this is reasonable.

Your loving son

Freddy

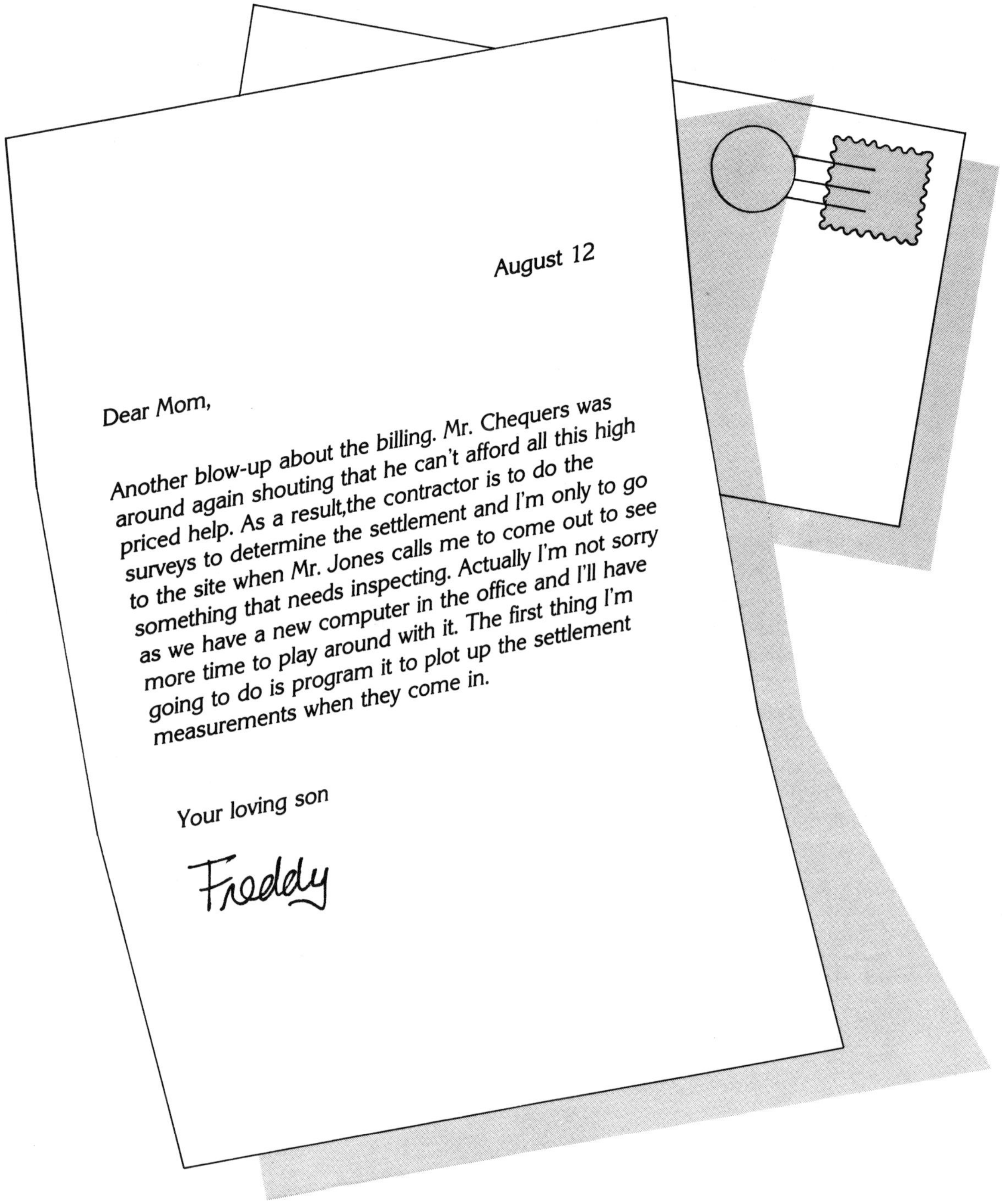

August 12

Dear Mom,

Another blow-up about the billing. Mr. Chequers was around again shouting that he can't afford all this high priced help. As a result,the contractor is to do the surveys to determine the settlement and I'm only to go to the site when Mr. Jones calls me to come out to see something that needs inspecting. Actually I'm not sorry as we have a new computer in the office and I'll have more time to play around with it. The first thing I'm going to do is program it to plot up the settlement measurements when they come in.

Your loving son

Freddy

August 19

Dear Mom,

When the first readings came in from the site there were only three records so I phoned Mr. Jones to see what had become of the other five. He said a truck had run over the other stakes. When I told Mr. Hardwood he really hit the roof. He's normally a quiet easygoing person but this time he called Mr. Jones and told him in no uncertain terms to replace those stakes or he wouldn't be responsible for approving the site.

Your loving son

Freddy

Sept. 5

Dear Mom,

Things have been going along very nicely. Last week I had a few days off and combined with the long weekend I went off fishing with Bob and some of the other lads. Bob tells me that Mr. Jones's brother-in-law is the chief accountant for the paint company and that the company actually owns the contracting company.

Your loving son

Freddy

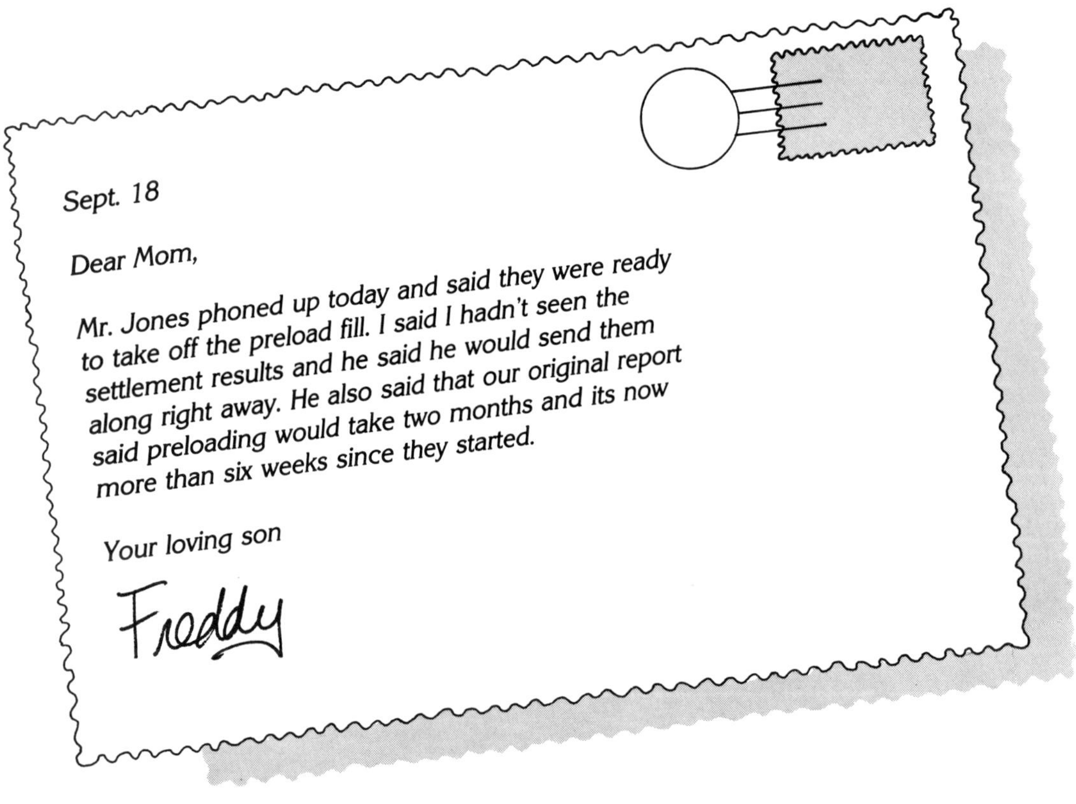

Sept. 18

Dear Mom,

Mr. Jones phoned up today and said they were ready to take off the preload fill. I said I hadn't seen the settlement results and he said he would send them along right away. He also said that our original report said preloading would take two months and its now more than six weeks since they started.

Your loving son

Freddy

my parka and long johns. I'll probably be there until after Christmas as they are anxious to get this job finished before the end of the year. Something to do with government grants or taxes which run out December 31.

Your loving son

Freddy

Sept. 30

Dear Mom,

When the settlement results arrived last week only one of the original hubs was in place. Three of the hubs hadn't settled since the previous week and one had actually risen almost 3 inches. I can tell you it was difficult to make much sense out of the plots that the computer produced. Mr. Jones phoned back about an hour after he delivered the survey notes and asked if he could remove the preload. I told him three of the hubs showed no change from the previous week and he said that was good enough for him.

They must have worked over the weekend because on Monday he phoned for me to send someone out to inspect the footings because they want to pour concrete that afternoon.

Mr. Hardwood has been away at the annual meeting of professional engineers. I told him what had happened and he agreed that it was too late to do anything and that it would probably be ok.

We have another job supervising a small dam in the north and I'll probably be going up there. Please send

June 15

Dear Mom,

I think I told you awhile ago that there was a good deal of unhappiness about that job I was on a year ago. After I went up north they built the warehouse and the paved floor broke up so that the loaders couldn't move the barrels of chemical around. Anyway today I went with Mr. Hardwood to see the lawyers who are looking after the insurance claims which Mr. Chequers is making. Their offices are up at the top of one of those big downtown buildings and you get a terrific view. Just like being in a helicopter only not so noisy. Mr. Sharpy and Mr. Pinstripe asked all about what I remember of things that went on a year ago. They said they were on our side in this but they didn't sound very happy when I told them about the fill being lower than planned or about the survey hubs being replaced. Anyways I'm sure glad Crossgain and Associates had insurance as it protects employees (me) as well as everyone else.

Your loving son

Freddy

Drain in a silty sand which did not function properly due to lack of a filter to match drain pipe openings and natural soil; also failure to recognize significant iron content in groundwater causing reduction of pipe openings with time.

A geotechnical engineer should not take responsibility for a drainage system unless he has been specifically involved in the detailed design and follow- up inspection services during construction.

18

Lawyers, Judges and Expert Witnesses

Lord Justice McKinnon, speaking of the Trade Marks Act: "I have very little notion what the section is intended to convey, and particularly the sentence of 253 words which constitutes subsection 1. I doubt if the entire statute book could be searched for a sentence of equal length which is of such fuliginous obscurity."

Fuliginous means sooty

A practising geotechnical engineer may have no ambition to appear in court as an expert witness. Nevertheless if he or his firm are involved in a professional liability claim he will probably have to appear as a witness on technical matters and may be asked to assist his legal counsel in evaluating evidence presented by expert witnesses hired by the opposing side in the dispute. He will have to give his lawyer a crash course in his technical specialty and should expect to receive a crash course in courtroom procedures.

Law and engineering are both problem oriented professions and people attracted into these fields are people who enjoy solving problems. For each problem there is a best type of approach to problem solving and in many cases a problem can only be solved by using specific tools and techniques. **Problem solving in engineering is principally by means of numerical and graphical procedures while problem solving in law is almost entirely by means of words.**

Problem solving in engineering is principally by means of numerical and graphical procedures while problem solving in law is almost entirely by means of words.

This difference in the kinds of problems to be solved and the pertinent problem solving techniques (words vs numbers) leads to difficulties in communication between lawyers and engineers. Lawyers who enjoy their profession and are successful at it, are problem solving wordsmiths. Successful engineers have the ability to think in terms of numerical models of materials, space and forces. Frequently the two do not meet on any common ground. Lawyers do not always recognize the existence of engineering problems and most engineers have a great distrust of language as a valid problem solving tool.

In Canadian courts professional liability cases are tried before a judge. This judge is appointed to the bench from the legal profession and accordingly is by inclination, training and experience a language oriented problem solver. There are a number of reasons why appointment as a judge is an attractive step for many lawyers. One of these is the variety of cases which he will hear. A successful lawyer in a large firm may get into a routine of some special field such as divorce or corporate law so that the same problems are repeated throughout the year. The judge, however, enjoys a variety of cases, divorce today, fraud next week followed by crime and a professional liability case.

The problem of the geotechnical engineer involved in a professional liability claim can be clearly seen. He must first successfully communicate with his legal counsel who by temperament, training and (you hope) experience is skilled in the law which probably doesn't contain one equation in a hundred volumes. The lawyer must then guide the engineer in the presentation of his side of the case to a judge who may have had no previous contact with the concepts and practice of this particular branch of engineering. It is a very demanding educational process for all concerned.

Courts recognize that a person who is by training and experience an expert is better qualified than a lay person to arrive at a correct opinion based on facts in his field of expertise.

One of the anecdotes recounted by Lord Alwyn Jones in an address to the College of Law at the University of Saskatchewan concerned a Justice of the Court of Appeal who complained to a solicitor who was explaining his case in great detail, "You must give us credit for knowing something". Counsel replied, "That was the mistake I made in the lower court, my Lord".

This problem of communication is further aggravated by the fact that the judge and the engineer are comfortable working in what are two different media, language and mathematics. At a recent seminar on alternative careers in law at the University of Victoria one of the speakers identified his job as that of translating engineering and scientific evidence into English that would be comprehensible to the court.

The objective of the courtroom procedure is to present to the judge the facts of the case and remind him of the applicable law. The judge, based on the facts which are presented, forms his opinion as the basis of his judgement. The legal system assumes that the judge is capable of forming a correct opinion based on the facts. Witnesses are therefore limited to presenting facts and are prohibited from expressing opinions since this would usurp the responsibility of the judge to form an opinion based on facts.

One of the few exceptions to this rule is the expert witness who may express an opinion based on facts which have been established, or on a hypothetical set of facts which are presented to him. The reason for this exception is that the **courts recognize that a person who is by training and experience an expert is better qualified than a lay person to arrive at a correct opinion based on facts in his field of expertise.** Thus an expert in ballistics is allowed to express an opinion as to whether or not a bullet was fired from a certain gun based on his examination of the bullet and the gun. If only the facts on which he based his opinion were presented, scratches on the bullet and condition of the gun barrel, the judge or jury could not form a valid opinion. Nevertheless the expert's opinion is still an opinion and may be ignored by the court if they so wish.

In general the engineering profession does not present the united front against professional liability claims that is often ascribed to the medical profession. Accordingly, a lawyer advancing a professional liability claim can usually find an "expert" to support his case against a geotechnical engineer. The geotechnical engineer will accordingly be required to assist his lawyer in evaluating, and if possible discrediting, the evidence of the opposing expert.

One of the first things to keep in mind is that the judge may completely ignore the opinion of an expert witness. The extent to which the judge accepts the expert's opinion may depend more on the judge's perception of the expert's qualifications and presentation than on the validity of the expert's reasoning.

It is generally agreed that the most effective expert witness appears impartial, authoritative, and logical.

It can be assumed that very few expert witnesses are completely impartial and unbiased.

It can be assumed that very few expert witnesses are completely impartial and unbiased. He has been employed by the lawyer who is presenting a compendium of facts and opinions to support his client's case. The lawyer's questioning is intended to lead the expert through this evidence so as to highlight the favourable aspects and minimize the unfavourable. If the expert is biased toward his client he will endeavour to continue this process in his replies to cross examination but he must be very cautious because nothing is more damaging to his credibility than to allow his bias to show. If the expert is truly impartial he will be much more willing to admit the validity of an alternative explanation or opinion if in fact there is one.

From the point of view of the expert's lawyer the ideal expert is one who appears impartial but who is in fact as committed to the client's case as the lawyer himself.

The authority by which an expert presents his opinions is usually indicated at the outset of his testimony by a recitation of his experience, education, publications and awards. If the opposing expert's testimony is crucial it may be useful to check his qualifications and run a library search of his publications.

The expert will present his opinions in an authoritative manner but there is a difference between authoritative and dogmatic. If his words or manner imply "I am the expert. I have considered these facts. This is my opinion and you must accept it ", he may be vulnerable to questioning on the reasoning he followed from the facts to his opinion. He may not be able to reproduce the line of reasoning or the reasoning may include some steps which the judge may find illogical or at least fuzzy.

Lay people tend to accept scientific and engineering conclusions as definite and believe that the processes by which they are reached are also definite and straightforward. They may be uneasy when they see the convoluted processes by which the conclusions are reached. An assistant to a well known chemist who frequently appeared as an expert witness in criminal cases commented that if he were the accused he would merely present a photograph of the expert's laboratory. No jury would believe that valid results could be produced from the chaos of notes, glassware, samples and reagents.

Many experts feel that they are qualified as experts not only in their own field but also in all fields. They may be encouraged in this tendency by their own lawyer. It has been said that an expert has more to fear from his own lawyer who will push him into making statements he cannot support than from the opposing counsel. The field of geotechnical engineering embraces laboratory testing, analytical procedures, engineering geology, rock mechanics, field testing, hydrogeology and many others. Few practitioners are truly experts in all of these fields and fewer still are also experts in the broader fields of civil engineering.

An expert witness should be thoroughly prepared by his lawyer so that he is aware of the significance of his testimony in relation to the testimony of

other witnesses and the case as a whole. Expert witnesses are paid significant fees and a client may be tempted to minimize his expense by bringing in the expert without adequate preparation. If the expert isn't well prepared the opposing lawyer may take advantage of the opportunity to show that the expert is not aware of the significance of his testimony.

An expert witness should be thoroughly prepared by his lawyer so that he is aware of the significance of his testimony in relation to the testimony of other witnesses and the case as a whole.

An expert witness is asked to give his expert opinion based on certain facts. These may be facts which he himself has observed and which he presented as evidence. They may be facts observed by others and presented in evidence by other witnesses. They may be hypothetical facts presented by the lawyer as part of a question. The expert must accept these facts and form his opinion based on them. In evaluating the expert's evidence it is important to clearly identify the source and character of the facts on which he bases his opinion.

 Mephisto's UNETHICAL ADVICE

A regular feature of the Devil's Advocate and the Hades Daily News

Dear Mephisto:

I'm an engineer with an MBA and I know what things cost. That's the reason I'm in charge of this new building for the Department of Inferior Products. The minister says that the building has to be finished by next March and my firm Rush, Forth and Doitt have been appointed managers to fast track this production. The site has been chosen, the architect has conceptual plans the structural engineer is appointed, the contractor is moving dirt and the geotechnical engineer has written a report.

Every Monday morning about ten people turn up at my office for three hours for a review of the project. Each of those professionals costs $75 to $100 an hour. I figure that each of these meetings costs at least $2000. The one that gripes me the most is the geotechnical engineer. For example, today we were discussing critical path planning and since foundation construction comes first, he had nothing to say after the first five minutes.

How can I keep these costs under control?

Yours

M.I. Haste

Dear Haste:

The solution is obvious: cut down on the meetings. Once every two weeks is plenty. And cut down on the people attending. All you really need to make decisions are the architect, the contractor and yourself. The architect can tell the structures man everything he needs to know -; he can also read the soils report and interpret it for any decision that needs to be made. For that matter you could read it yourself. You went to university didn't you?

Best of luck.

Mephisto

Mephisto's UNETHICAL ADVICE

A regular feature of the Devil's Advocate and the Hades Daily News

Dear Mephisto:

I'm planning on building on the outskirts of town and asked a soils firm to drill some holes on the site. The letter I got back tells what the engineer will do and how much it will cost, $2500 and ends with this statement "I hereby authorize the work described in the foregoing letter and will be personally responsible for payment of invoices submitted.

Mary Jane Developments per
PRESIDENT

The question I have is, will I have to pay his bill if the project doesn't go ahead?". I might say I'm hoping to get my cousin to invest in this and so far haven't been able to arrange a mortgage or get any firm commitments to rent the stores.

Yours truly,

I.M. Optimist

Dear Optimist:

The answer to your question is "possibly" - it really depends on how determined the engineer is to collect his fee. I assume you have already transferred your house and bank account to your wife's name.

What you have encountered is a soils engineer who doesn't want to finance your project and you should probably look for another consultant. One alternative would be to phone him up and say "Go ahead, we need the report by Friday and because $2500 is too much cut the work back to only half the number of boreholes". Then when he tries to collect his bill you won't be personally liable and there will be enough uncertainty as to what was authorized that you can probably beat him down on the amount of the bill. If he writes a report or does any work at all you can claim that the work was incomplete and recover something from his insurance company.

I don't have much hope that he would go for this. You need a less experienced engineer.

Good luck,let me know how it turns out.

Mephisto

19
Case History IX

This case history is copied almost word for word from the written judgement of the trial judge who heard the case. Where the original judgement gives names of those involved the appropriate terms Contractor, Owner, Engineer, Technician, etc. have been substituted. Although longer than some of the other cases it is valuable because it is clearly written and permits the reader to follow the reasoning by which the judge arrived at his decision.

The defendant is a one-engineer soils engineering firm against which the plaintiff seeks to recover for the failure of a concrete floor in a warehouse on its land which settled because of inadequacies in the design and application of a"preload" of piled sand which had been used to compress the peaty soil in preparation for construction.

The soils engineering firm (which I shall refer to as the "the defendant") was not engaged to design or supervise the preload. Nor was it given the information which it would require in order to express an opinion on the appropriateness of the preloading which the plaintiff did. But the plaintiff says the defendant, though not retained or paid to advise on the matter ought to have known that it was being relied on for advice and had both a duty to take care not to mislead the plaintiff and its contractors and a duty also to warn of danger which it should have foreseen in what the plaintiff was doing.

The present action as originally framed was also against the contractor who constructed the plaintiff's building, but this claim was settled before trial. By agreement between the present parties the court was asked to determine the percentage share of fault, if any, properly attributable to the defendant — as opposed to that attributable to the plaintiff and the contractor for the settlement and failure of the floor, and to assess the amount of damages, if any, properly chargeable to the defendant on the basis of any such apportionment of fault. The terms on which the claim against the contractor had been settled were not disclosed.

In order to assist the reader the following is the cast of characters

Mr. Doe - (plaintiff) owner of Blank Developments

Mr. Smith - (defendant) owner of one-man geotechnical firm

Mr. Jones - technician employed by Mr. Smith

Mr. Brown - engineer for the contractor

The Background

The plaintiff, Blank Developments Ltd., is a company belonging to Mr. John Doe and a partner whose affairs, at least in the context of the project in question, have been managed by Mr. Doe.

The building site was acquired for the company in 1975. The intention was to build a warehouse or workshop building there, and to rent out space in the building to a tenant or tenants engaged in light industrial or commercial businesses. Neither Mr. Doe nor his partner had any significant previous experience in construction. They made enquiries about a cement-and-wood building, to be constructed on a wholly-piled foundation, but found the cost — in the neighbourhood of $300,000 — too high for the project to pay its way. During 1978 they heard of building alternatives which might make the project economically feasible. They also found out something of the dangers inherent in the use of less costly methods of foundation design.

Early that year Mr. Doe learned that there had been settlement in the floor of a building on the next-door property, and also something of its cause.

This building had been constructed with a piled perimeter foundation supporting the walls and a cement floor "floated" inside on unpiled ground. This foundation design had been adopted against the recommendations of a soils engineering firm retained by the owner. Mr. Doe was shown the soils engineering report in question. Its most significant passage reads:

We understand that you intend to pile support the structure and were intending to "float" the floor. Based on the depth of peat encountered we do not recommend that the floor be supported by any means other than pile support. Site conditions such as these warrant total pile support for the building.

This reference to a "mixed" foundation is significant in the present context. It was this very technique which the plaintiff was ultimately to adopt for its own building. It was to do so with knowledge of the consequences which had flowed from the use of that design in the case of the building next door.

Some knowledge of the way in which "preloading" works is essential to an understanding of the problems which lay ahead for the plaintiff.

In such peaty soil conditions, preloading is generally a less expensive but more time-consuming method of foundation preparation than piling. Done carefully it will eliminate, or at least minimize, the risk of settlement taking place after a building has been erected on the prepared site. The compromise adopted for the neighouring building, and for which the plaintiff was to opt in the end, involves a pile-supported concrete perimeter foundation for the walls with a "floated" slab poured on preloaded soil inside. Engineering opinion is divided as to the wisdom of adopting this mixed foundation design. The evidence suggests that a relatively small settlement, which might be tolerable were the whole building on a "floating" slab, can play havoc if the walls are stabilized on piles and the floor alone is floated on unpiled preloaded soil.

The technique of preloading, while neither particularly complicated nor exclusively within the province of the soils engineer, calls for certain expert attention both in the planning stage and in application.

The amount of sand required for preloading a peaty soil must exceed by an appropriate margin the greatest weight which will subsequently be imposed on the ground which it is to compress. The preload is usually a sand pile shaped, very roughly, in this manner:

The crown of the pile has to extend beyond the boundaries of the actual building site, or "envelope". The sand must be uniformly shaped, so that the site will be uniformly compressed. The load must be kept in place until all settlement has ceased. In calculating the amount of preload applied the engineer must exclude any part of the material which is to be left on site to restore the original ground level after compression, or to raise it to a new elevation. That constitutes part of the weight which the soil must be prepared to carry, not part of the preload. The preload is that portion only of the added material which will be taken off the building envelope after settlement has ceased.

Thus the design of an appropriate preload requires calculation of the weight of the proposed building and contents, the weight of the material to be left in place as fill, and the weight of the material to be removed. The preload must be properly shaped and so placed that this crown overlaps the building envelope. Settlement must thereafter be completed — stability must be achieved before the preload portion of the material can safely be removed.

These are some, at least, of the matters to which the mind of an engineer must be directed in designing and supervising a preload.

When Mr. Doe was looking for an economic solution to his construction problem early in 1979 he must have known that a partially-piled foundation with a floated slab floor on preloaded grade would probably be cheaper than an all-piled foundation. He knew that such a design — in which the floor area only is floated — had been rejected for these soils conditions by one soils engineer, that it had been proceeded with notwithstanding that advice and that it had failed. While he knew very little about preload, Mr. Doe knew enough to recognize that he would need expert guidance in order to minimize the risks involved if that design should be adopted for his own project.

In May and June of 1979 Mr. Doe discussed his requirements with personnel from the Contractor who provides and erects pre-fabricated steel buildings, and was quoted more attractive prices.

The cast:

Mr. Doe - *(plaintiff) owner of Blank Developments*

Mr. Smith - *(defendant) owner of one-man geotechnical firm*

Mr. Jones - *technician employed by Mr. Smith*

Mr. Brown - *engineer for the contractor*

While these discussions with the Contractor originally centred around an all-piled foundation design, the Contractor also mentioned to him the possibility of a "floated" floor. Mr. Doe brought a quantity of sand onto the property and dumped it in individual truck-load piles within the building envelope. He says he did this not for the purpose of preloading, but with a view to raising the level of the site on which he intended to build.

During the discussions between Mr. Doe and the Contractor the representative of the Contractor said they would need to have a soils test done, and recommended that the defendant be asked to do it.

The Preliminary Report

The defendant is a company through which Mr. Smith carried on his practice as a soils engineer with the assistance of three employees — two technicians and a secretary.

It was one of the technicians, Mr. Jones, who answered a telephone call from Mr. Doe on June 12. Mr. Doe described the sort of building he had in mind and said he was planning to build on an all-piled foundation. Mr. Jones said that was a good idea in view of the soils conditions in the area. Mr. Doe said that he was thinking of having the Contractor erect the building and that they needed a soil investigation. He mentioned that the owner of a nearby building had experienced settlement problems. Mr. Jones suggested a three-hole test program as appropriate and said he would get a driller to quote a price and let Mr. Doe know the total cost. After getting the drilling quotation, he phoned back and said the cost would be $900. Mr. Doe phoned later and said they didn't want to pay that much; he asked for something less elaborate. After discussion with Mr. Smith, Mr. Jones quoted $400 for a report on a single test hole, and Mr. Doe accepted.

Mr. Jones went to the site three days later and supervised the test. The nature of the test and the conclusions which the defendant drew from it are described in a document dated June 18, which plays a central role in the present litigation. Headed Report of Preliminary Subsurface Soil Investigation and Recommendations, it reads as follows:

Introduction

In accordance with your request a preliminary subsurface soil investigation was conducted June 15 at the above project site. The proposed 50 foot by 100 foot building will be steel frame with metal siding. A pile foundation is planned. This report presents recommendations for the pile support of the foundation and for the slab-on-grade floor.

Investigation

One penetration test hole was placed at the location shown on the attached Test Location Plan. A modified top drive Mayhew drill rig was used to a depth of 50 feet. A 4 1/2 inch diameter auger hole was bored to 12 foot depth to explore the upper soil strata. This hole was placed about 3 feet south of the penetration hole.

Description of Site

The site is uniformly flat. No trees exist. Stockpiles of river sand have been deposited on the building site to a depth of about 9 feet for the purpose of preloading the slab area.

Description of Subsoil

The upper 20 feet of the site is composed of a brown non-fibrous peat with some fibrous peat mixed in. This material is soft and saturated below about 6 foot depth. The peat is mixed with clay from about 20 to 30 feet and probably changes at about 30 feet to a sand and silt which exists to the maximum 50 foot depth explored. This sand provides suitable bearing for piles.

Conclusions and Recommendations

The upper 30 feet of soil is unstable and will consolidate under anticipated floor loads. Preloading is advised to stabilize this soil. The river sand currently on site is suitable preloading material. Use 1 foot of this sand as surcharge for each 95 psf of dead and live load anticipated on this floor. The penetration test indicates that individual size 13 piles (minimum) driven to 50 foot depth will develop 10 tons Allowable Bearing Capacity. The same size piles driven to 60 foot depth may develop an Allowable Bearing Capacity of 20 tons if the sand density increases however this investigation terminated at 50 feet and this increase in density was not substantiated. The piles may be either: used, marine piles, 10 pcf creosoted foundation piles; or green pile for the lower section and creosoted 15 foot top section. A securely fastened pipe splice is recommended to join the upper and lower sections of the 2 piece pile. The pile driving operation should be supervised by someone competent in this type of work in order to ensure adequate bearing for the piles on this project.

If questions should arise, please contact the undersigned or Mr. Jones.

B. Smith

Mr. Smith, P.Eng.

The position of the plaintiff is that this report gives the appearance of approving use of the sand there — as dumped in truck-load piles on the site — for a preload, and misled the plaintiff into following that course. The plaintiff says the references to the investigation as "preliminary" contained in the heading and opening sentence of the report, are not sufficient to constitute a warning that recommendations are not to be used for construction purposes.

The cast:

Mr. Doe - (plaintiff) owner of Blank Developments

Mr. Smith - (defendant) owner of one-man geotechnical firm

Mr. Jones - technician employed by Mr. Smith

Mr. Brown - engineer for the contractor

The report was put into final form and approved by Mr. Smith. It was picked up soon afterwards by Mr. Doe. Mr. Doe read the report, but he says he regarded it as something intended for the contractor rather than himself. So he took it to the contractor's office.

The contractor is a company which has professional engineers on its staff. It supplies and erects prefabricated buildings, with ancillary engineering services, including foundation design and site inspection. It is quite apparent that both Mr. Jones and Mr. Doe intended the soils report to be used by the contractor's engineering personnel, for whom it had been ordered. There is no suggestion that anyone thought it was intended for the guidance of laymen, such as Mr. Doe and his partner.

I have concluded that this report was intended to be "preliminary" in the sense that its purpose was to assist a construction engineer in costing, and deciding between, foundation alternatives. It was not intended to be used for actual foundation construction, though the information concerning the piled foundation was probably adequate for that purpose.

The Design Phase

Sometime during the latter half of June the plaintiff retained the Contractor to supply and construct the prefabricated building and to perform engineering services required for the project.

The Contractor was not to be a "general contractor", in the sense of having total responsibility for the whole work, and actual preparation of the site and foundation construction were specifically excluded from its contract. But the matter for which it undertook responsibility included, among others: "Foundation design including letter of supervision and site inspection" and "foundation design drawings, signed and sealed by a registered Professional Engineer. The plaintiff is said to have been "its own general contractor" in the sense that the plaintiff was to arrange, at its own cost, for all work required other than that undertaken by the contractor, including site preparation work and construction of the foundations. But as part of its lump-sum contract the contractor undertook to design the foundations and to inspect the site prior to construction. The contractor was to provide a supervising engineer for the project, in addition to providing and erecting the "pre-engineered" steel building.

The Contractor proceeded with the preparation of drawings. These contemplated in place of the all-piled foundation which had originally been planned, the less-expensive mixed design — a concrete foundation supported by piles for the walls and concrete slab floor poured on preloaded, unpiled ground inside. Before completing these drawings, the contractor's chief engineer telephoned Mr. Jones, the defendant's technician, to ask about preloading

The contractor's engineer, Mr. Brown, asked Mr. Jones how long the preload should be left in place. He says Mr. Jones replied that it should remain in place for eight weeks or until settlement ceased. Mr. Jones says he replied that he did not know how long settlement would take, that he had heard reports of eight weeks being a sufficient time for settlement to take place, but that the way to find out was to use settlement gauges. Mr. Brown told Mr. Jones he was going to make some reference to preloading in the drawings, but he did not indicate what it was he intended to put on the plan.

I found Mr. Jones a credible witness and his recollection of this conversation seemed somewhat better than that of Mr. Brown.

Following this conversation Mr. Brown put the following, in capital letters, as a note to the Foundation Plan drawings:

SITE TO BE PRELOADED AS PER SMITH'S LABORATORY REPORT DATED JUNE 18, 1979. PRELOAD TO REMAIN IN PLACE 8 WEEKS OR UNTIL SETTLEMENT CEASES.

Thus it was that part of the defendant's preliminary soils investigation report and some of Mr. Jones remarks on the telephone to Mr. Brown concerning settlement time became transformed into specific construction specifications on the final foundation plan. I am satisfied that the defendant's personnel never approved of this notation, and that they remained unaware of it until after construction had been completed and the settlement had occurred.

Mr. Brown was candid in conceding his responsibility for the project, and that there had been some oversight on the Contractor's part.

He said he knew that settlement gauges must be used in order to be sure when settlement has ceased, that the preload should not be removed until settlement has ceased, and that a preload must extend beyond the edges of the actual building envelope. He said he assumed from the defendant's report that a properly designed preload was then already in place. While the report was described as "preliminary", it did not seem preliminary in substance, he said, because it contained specific recommendations and conclusions. He thought the defendant would have known, as a result of his telephone conversation with Mr. Jones, that the report was being used for design purposes. He said he assumed from this conversation that monitoring the preload would not be necessary if it remained in place for eight weeks.

While maintaining that the Contractor was not qualified to design a preload, Mr. Brown agreed that it had the responsibility to see that a proper foundation design was done, including the preload, and to inspect the site. No preload design was in fact provided to the plaintiff, nor did the Contractor or any engineer on its behalf, inspect the site before building commenced.

Mr. Doe testified that he received the plans from the contractor on

The cast:

Mr. Doe *- (plaintiff) owner of Blank Developments*

Mr. Smith *- (defendant) owner of one-man geotechnical firm*

Mr. Jones *- technician employed by Mr. Smith*

Mr. Brown *- engineer for the contractor*

June 27 and read the note reproduced above. He understood it to mean that the individual truckload piles of sand dumped within the building envelope constituted a proper preload, and that all he had to do was leave them there for eight weeks and the ground would be ready to support the floor.

Mr. Doe said he assumed that the Contractor had drawn correct conclusions from the defendant's report. For that reason, he said, he put reliance on the report, and did what he thought it said.

Should the defendant then, in preparing its report, have foreseen the possibility that this might happen?

Mr. Smith believed, quite correctly, that there was an engineer in overall charge of the project. His firm was retained in a very limited way to do a basic soils test for $400. He identified the investigation as preliminary only. He assumed that the site was to be inspected by the supervising engineer, and that the engineer would have some knowledge about preloading. The report said there were "stockpiles of river sand deposited on the building site", and that this sand "constituted suitable preloading material". Its only recommendation on preloading was "use one foot of this sand as surcharge for each 95 psf of dead and live load anticipated on this floor". It did not say there was a preload in place, nor did it say how to create or employ one. The report said only that there was suitable material on site and how much would be needed.

I have concluded that the defendant could not have been expected to foresee the possibility that an engineer in charge would refer to this report as an instruction on preloading, or that he would interpret it to mean that a properly-designed preload was in place, and not make any inspection. The report says nothing about the shape or position of the preload, or how to know when to remove it. Nor is there mention of the volume which must be removed, as opposed to that to be left in place as fill.

Should remarks made by Mr. Brown, then, in his telephone conversation with Mr. Jones, have alerted the defendant to the possibility that the report, or Mr. Jones comments, might be used by Mr. Brown as they were?

Mr. Jones could not, I think, have guessed, without being told, that the report was being relied on as indicating that a properly-designed preload was already in place. Nor do I think he could reasonably be expected to foresee that the Contractor intended to put on the plan a note capable of being interpreted as meaning that monitoring of settlement was unnecessary — that merely leaving a preload in place for eight weeks would be sufficient to assure that the necessary settlement had occurred. I say that particularly because I accept that Mr. Jones did mention the need to monitor in his conversation.

I cannot therefore find that there was negligence on the part of the defendant up to this point, which, had the defendant's involvement then ceased, could be said to have contributed to the ultimate failure.

The Pile-Driving Inspection

During July Mr. Doe retained a pile-driving company to put in perimeter piles in accordance with the contractor's foundation plan, and asked the defendant to send someone to supervise the operation.

Mr. Jones attended at the site for this purpose July 12 and 13. He found the sand on the site had been arranged so as to make room for the pile driving crew to do that work around the perimeter. Mr. Doe spoke to Mr. Jones about the preload during the course of the pile-driving, and he drew Mr. Jones' attention to the failure of the building next door. He testified in court that he asked Mr. Jones how the preload seemed and that Mr. Jones replied that it was "fine" and to leave it on for eight weeks. In cross-examination Mr. Doe said that this answer was given in a "off-hand" way, but that he relied on it. He said he relied also on the defendant's report in coming to the conclusion that preloading was being properly done. He denied that Mr. Jones mentioned the use of settlement gauges during this brief exchange.

On examination for discovery, Mr. Doe had said he relied solely on what he was told by Mr. Jones on this occasion, so far as the preloading was concerned, and not on anything contained in the report. Mr. Doe also said on discovery he had understood, until he spoke to Mr. Jones on this occasion, that he would have to keep the preload in place for longer than eight weeks if settlement had not ceased when the eight weeks was up.

Mr.Jones' evidence was that Mr. Doe pointed at this meeting to the sand and asked what Mr. Jones thought of the preload and that he answered that it seemed high enough. He said he asked Mr. Doe how long it had been on and Mr. Doe indicated about three months and asked if that was long enough. He said he replied that it might be but that one would have to use settlement gauges to be sure. He said Mr. Doe asked if that was really necessary and Mr. Jones replied that Mr. Smith always used them.

I accept Mr. Jones' evidence as a reasonably accurate account of the exchange which took place between them that day.

I cannot find that this casual conversation should have caused Mr. Jones to realize he was being relied on to warn the plaintiff of any inadequacy there might be in the preloading procedure. His firm had not, of course, been retained for that purpose and his visit had nothing to do with it. The approval which he expressed in an off-hand way could only have related to the quantity of sand. It could not have related to the configuration — which had in any event been disturbed for the pile-driving — nor to whether settlement had ceased. I accept Mr. Jones' evidence that he told Mr. Doe that settlement

The cast:

Mr. Doe - ***(plaintiff) owner of Blank Developments***

Mr. Smith - ***(defendant) owner of one-man geotechnical firm***

Mr. Jones - ***technician employed by Mr. Smith***

Mr. Brown - ***engineer for the contractor***

must be monitored in order to know whether it had ceased.

Mr. Doe's prior understanding, as expressed on examination for discovery, that he would have to leave the preload in place if settlement was still taking place when the eight-week period mentioned on the foundation drawings expired, was a manifestly reasonable one which could not reasonably have been displaced by this conversation with Mr. Jones.

I cannot say that Mr. Jones was negligent in the remarks which he made to Mr. Doe on this occasion.

Despite his understanding that settlement had to be checked, Mr. Doe proceeded to level out the preload without knowing whether it was still settling. He spread the sand over the actual building envelope and the adjacent parking area so as not only to fill in inundations caused by the preloading but to raise the building envelope to a higher elevation. He then had the pile-supported cement perimeter foundations built and thereafter handed the job over to the contractor for construction of the building.

The only further involvement of the defendant during the foundation phase of the project was the conduct of laboratory tests on concrete and pile-cap samples provided to it. This did not involve work at the site.

Ought the defendant of its own volition to have volunteered a warning about the preload during this period?

The plaintiff says the defendant ought to have realized that Mr. Doe was inexperienced and that he might be proceeding on a dangerous course — that the preload probably had not been properly shaped, and was not being monitored — and should have given him a warning. The fact that the defendant had not been engaged for preload design or supervision is no answer the plaintiff says, to this allegation of negligence in failing to give some sort of warning during or after Mr. Jones' July 12-13 visit.

When he returned from the pile-driving operation, Mr. Jones told Mr. Smith of his conversation with Mr. Doe and said that he saw no settlement gauges. There can be no doubt that Mr. Smith, had he been asked to give his advice in the matter at this stage, ought to have expresed doubts on whether the preloading had been competently done. He had himself been at the site briefly during each of Mr. Jones' visits. On neither occasion was there a properly-shaped or properly-positioned preload, although this could on both occasions to some extent be explained. He had no knowledge that there was a proper preload in place, and good grounds for doubting it. He knew from Mr. Jones that it was unlikely settlement was being monitored.

Mr. Smith testified that as a consultant with a strictly limited engagement he had no justification for involving himself.

He had, of course, been retained for restricted purposes. He knew there was a supervising engineer in charge and had confidence in that engineer's ability. A supervising engineer is taken to accept responsibility for all necessary engineering functions which have not been delegated to others. Had he been in Mr. Brown's shoes, Mr. Smith said, he would not have appreciated gratuitous interference from a soils consultant in a matter in which the consultant had not been engaged. Mr. Smith felt that the Contractor had chosen either to use its own resources or take advice elsewhere with respect to the preloading, and it did not seem to him that he could properly involve himself in the matter.

I have no doubt that there are circumstances in which a professional man may have a duty to warn in connection with matters about which he has not specifically been engaged. But where he knows that another member of his calling has been retained in a matter it is difficult to conceive of such circumstances — short, in any event, of those involving hazard to life — in which he would be under a duty to involve himself without first receiving a formal request for his opinion. The casual enquiry made of Mr. Jones by Mr. Doe seems to me to have fallen far short of what an engineer could regard as such a request.

I cannot therefore find that the defendant was at this stage, under a duty to make gratuitous enquiries, to offer gratuitous advice, or to warn the plaintiff of any risk to which it might be exposed.

The Inspector's Warning

The last on-site investigation, conducted by the defendant, occurred during September, was directly related to the possibility of settlement, and resulted from an expression of concern by the municipal inspector that the plaintiff's building might suffer the same fate as that next door.

The municipal building inspector suggested to Mr. Brown, the contractor's chief engineer, that he should satisfy himself that his design would not result in the sort of settlement which had occurred in the building on the adjoining property. At this point the shell of the plaintiff's building was largely completed, with the roof in place, but the concrete floor had not yet been poured. Mr. Brown telephoned Mr. Smith to pass on the message. He asked Mr. Smith if he would visit the site and see if there was any reason for such concern. Mr. Smith agreed and said he would call Mr. Brown back if he concluded there was.

Mr. Smith looked at the two buildings and took some photos. He concluded there was nothing to suggest that the plaintiff's building might be in any danger. He decided there was no reason to call Mr. Brown.

The reason Mr. Smith concluded that there was no need for concern

The cast:

***Mr. Doe** - (plaintiff) owner of Blank Developments*

***Mr. Smith** - (defendant) owner of one-man geotechnical firm*

***Mr. Jones** - technician employed by Mr. Smith*

***Mr. Brown** - engineer for the contractor*

was because he saw no visible evidence of distress in the case of the plaintiff's building. In the case of the building next door, on the other hand, there were obvious signs of settlement. There was a gap, clearly visible from the outside, between the ground and the pile-supported concrete perimeter wall foundations. There was also clear evidence of settlement of the floor inside. No settlement was evident at the perimeter of the plaintiff's building, and it had as yet no floor. Finding no similarity in the condition of the two buildings, Mr. Smith concluded that there was no need for concern.

Mr. Smith seems to have viewed his task on this occasion as that of an observer. He does not seem to have felt that it was his duty to make enquiries. I must ask whether he was justified in adopting this view.

It seems to me unlikely there could be evidence of settlement at that point on the plaintiff's site, even if the foundation conditions there were as defective as those next door. Only a month had passed since the sand had been spread out and the pile-supported perimeter foundations built. Since the floor slab was not yet poured, no weight had yet been imposed on the newly-created grade. The next-door building, on the other hand, had been completed and in use for more than a year. The preloaded soils there had long been under sustained stress, while the plaintiff's foundations had yet to be tested.

I have concluded that a visual inspection could do little, in these circumstances, to answer the question which the inspector had posed and which Mr. Brown had passed on to Mr. Smith.

In the light of what he knew and did not know about the preloading, and of what he ought as a consequence to have questioned, I think Mr. Smith had a duty to make enquiries before he could justifiably say that the plaintiff's building would not suffer the same fate as its neighbour. I think he had to know what sort of preloading was done in each case; certainly he had to know what sort of preloading had been done on the plaintiff's site. If he did not wish to pursue the matter beyond a visual inspection I think he was bound to tell the contractor that he could not answer the question put to him.

By his silence Mr. Smith implied that there was no need for concern. In an engineer having that special expertise and with the knowledge which he did have, to be silent in the circumstances seems to me to fall short of the appropriate standard of care.

I find there was negligent conduct also on the part of the contractor in failing to communicate to Mr. Smith information which it had and which it ought to have realized Mr. Smith would need in order to answer the question it asked. Mr. Brown should have disclosed the preloading instructions appearing on the foundation plan; he should also have said that the contractor had not inspected the preload and did not know how it had been done.

While there seems to me clearly to have been negligence on the part of the plaintiff in the conduct of the preloading, I cannot say it contributed to the defendant's failure to discover and warn of the danger following the building inspector's enquiry. That, I conclude, was due in equal parts to the negligence of the contractor and the defendant. But for their negligence, I find that the plaintiff would have been warned of the grave danger in proceeding with the floor slab, and would have taken remedial action.

Corrective measures which would have been instituted at that stage would necessarily have been less costly than those which had in the end to be undertaken after settlement occurred.

Conclusion

The only negligence of the defendant which I find to have been proved is in its misrepresentation by silence following the specific engagement by the contractor in September to advise on risk of settlement.

I have found the defendant was responsible as a consequence for 50% of the damages suffered because of the delay in remedial action between September, and the time when settlmeent took place. I have reached that conclusion because I find: (a) that at no time prior to September, 1979, did the defendant have reason to believe it was being relied on for professional advice as to the design, application, monitoring or removal of the preload; (b) that at no time prior to September, 1979, did the defendant give any advice on preloading which, properly considered, could have misled the person for whom it was intended; and (c) that no duty to warn rested on the defendant prior to September, 1979, because until that point it had not been engaged to give preloading advice and knew that another engineer was in charge.

I cannot say that use of the "mixed" foundation design was in itself contrary to competent engineering practice, even though it is plain that some engineers would have recommended against it. The evidence suggests that the system is one which, with competent design and application, could on this project have achieved a satisfactory result.

The reason the floor failed in this case was that the preload had been improperly shaped, irregular in height, not properly positioned over the building envelope and only partly removed, and perhaps also because settlement had not been satisfactorily completed.

It seems to me that the plaintiff and the contractor may have wished to avoid incurring the cost of obtaining preloading advice formally from a soils engineer. The enquiries made of the defendant by the contractor and Mr. Doe in June and July seem to have been carefully calculated not to assign responsibility to the defendant for the preload. Those enquiries may well have been cast in an informal way in order

The cast:

Mr. Doe - (plaintiff) owner of Blank Developments

Mr. Smith - (defendant) owner of one-man geotechnical firm

Mr. Jones - technician employed by Mr. Smith

Mr. Brown - engineer for the contractor

to avoid such a commitment as would justify a charge. I do not think a professional man can be made responsible for the work of others by carefully limited enquiry, or mere casual reference. Nor, I think, can it be expected that he will always hedge gratuitous responses to such informal enquiries by disclaiming responsibility.

The parties have agreed that the court should deal with apportionment of liability under the Negligence Act, R.S.B.C. 1979 Chapter 298, whether it be in contract or tort. It seems to follow that I need not be concerned whether the duty which I have found to be breached arose out of the original contract between plaintiff and defendant or out of the general law of negligence, nor do I think I need consider whether the duty breached by the defendant was one in respect of which the contractor might have claimed against it for indemnity or contribution, rather than a duty owed directly to the plaintiff.

Judgement

The total damages claimed are $98,667, consisting of $65,708 as cost of repair, and $8,251 as lost tenants' contributions to municipal taxes and $24,708 as lost rental income during the period for which the premises were incapable of occupation because of settlement and repair work.

I have endeavoured to segregate the costs and losses which would probably have been saved or prevented had the problem been identified before the floor had been poured and the interior of the building completed. I approximate the avoidable repair costs at $34,000, and I would attribute one-half of the 14-month period of rental loss to this delay in identifying the problem- in representing about $12,000 in lost income — and also one half of the resulting additional municipal tax burden, that is to say about $4,000. The delay in identifying the problem which was to result in failure accounts for roughly half of the total damages attributed to the failure. Thus I have said that the defendant is equally responsible with the Contractor for that additional cost and loss, or for about 25 percent of the total damages claimed.

I would have allocated the total fault for the failure of the building and resulting damages 25 percent against the defendant, 25 percent against the plaintiff and 50 percent against the Contractor and would have allocated the costs of the action in the same proportions. I find the plaintiff to be entitled as against the defendant to 25 percent of the total damages, or $24,667, plus 25 percent of the plaintiff's taxable loss, less 75 percent of the defendant's taxable costs. The plaintiff is entitled in addition to pre-judgement interest, at the rates awarded from time to time during the relevant period by the District Registrar on default judgements. The relevant period may, I think, fairly be set as being from March 1, 1980, to today's date as to one-half of the damages, and from September 1, 1980 to today's date as to the remainder.

Since the allocation of the damaged items which I have made was not addressed in argument, I would be glad to hear counsel if either party should feel that I have erred in apportioning the loss and expense incurred as between that which would and would not have been experienced had remedial measures been taken earlier, or in any other matter of calculation.

June 6, 1983

Where piling is recommended did the geotechnical report properly consider the combined effects of soil conditions, pile type and properties, proposed working load and required driving energy and termination set so that pile damage is not excessive? The ability to monitor installation and properly inspect deep foundations during construction, provides safeguards against defective performance.

20 Professional Liability Insurance

The first law of expert advice "DON'T ASK THE BARBER WHETHER YOU NEED A HAIRCUT"- quoted in The Official Rules

Insurance is a contract whereby the insurer undertakes to reimburse the insured in the event that he suffers a loss as defined in the insurance contract. In the case of professional liability insurance the loss is payment of an obligation imposed by law as a consequence of the actions of an engineer or architect in the course of his professional work.

When an owner or innocent third party suffers a loss, he and his lawyers will look around for some means of recouping that loss. There is no doubt that an engineer backed by a substantial professional liability insurance policy provides a more attractive target that an engineer with little or no assets. The plaintiff and his lawyers will go to great lengths to follow a tortuous path of reasoning to place some blame on a well insured defendant.

In this sense, there may be some logic to the feeling that you may be better off without insurance. However, very few geotechnical firms can operate without significant resources which could be seized to satisfy a legal judgement. Many clients insist that the consultants they hire are adequately insured and in some jurisdictions it is a condition of being licensed to provide consulting services that you carry insurance. Finally, a reputable firm will wish to shelter its professional and technical employees from personal lawsuits for liability arising from their professional work for the firm. Accordingly, most reputable firms carry professional liability insurance.

As noted in the introduction, professional liability claims against geotechnical engineers have been large and have risen rapidly over the past decade. The bulk of the claims are paid either by the insurers and/or by the geotechnical firms as their deductible portion of the policy. Consequently, professional liability insurance is expensive and the premiums form a significant portion of the annual overhead of most geotechnical firms.

Professional liability insurance is "catastrophe insurance" intended to protect the firm from a claim which would effectively ruin the business. The decisions regarding professional liability insurance are difficult business decisions which involve compromises between costs (direct and contingent) and risk. They can only be made intelligently by the geotechnical engineer/businessman with informed input from his insurance broker.

Most professional liability insurance is written on a "claims made" basis. This means the insurer will protect you against claims made while the insurance policy is in effect even though the activities which give rise to the claim occurred prior to the purchase of the policy, so long as you had no knowledge of an impending claim when the policy was purchased. The other side of this is that if the policy lapses, you are not protected against a claim made after the policy has lapsed even though the cause for the claim occurred while the policy was in force.

Professional liability insurance is "catastrophe insurance" intended to protect the firm from a claim which would effectively ruin the business.

Since claims may be instituted several, or even many years after the engineering activities which give rise to them, and since claims against geotechnical engineers are frequently complex and may take five or more years to settle, professional liability insurance, once entered into, is a long continuing relationship between the insurer and the insured. It is essential that you be comfortable with the insurer and his policies and that you are satisfied he has the resources and experience to provide this service for many years. Changing insurers is not to be undertaken lightly. It can be a serious problem for an engineer if his insurer should discontinue writing professional liability coverage.

Most professional liability policies cover all the activities of the firm and its present and past employees. Some policies are written to cover a specific project although these are usually set up to give joint coverage for a number of disciplines working on a large and complex project.

Most, but not all, professional liability policies, include coverage for the costs of legal defence against claims whether or not the claims are successful. The costs of the legal defence are not included in the deductible portion of the policy but are paid entirely by the insurance company. This encourages early reporting of possible claims. Early reporting is vital in minimizing the consequences of a claim. This is in the interests of the engineer who may require competent legal advice at the earliest possible stage of a potential claim. From the point of view of the lawyer, he must recognize that the interest of the insurer and the insured are not identical, particularly where the total amount of a claim may be less than the deductible portion of the policy or more than the total limit of coverage provided by the policy.

In order to lessen the cost of liability insurance, more policies are being sold in which the engineers deductible, or a portion of it pays the legal costs. Such a policy still obligates the engineer to report all circumstances of any potential claim. The insurer is still obligated to defend the engineer, with the engineer paying the legal costs.

There are a number of risk areas which are not covered by professional liability insurance. You should ensure that you are familiar with the terms of your policy in order to avoid inadvertently accepting risks for which you are not insured. A recent trend in many insurance policies, including professional liability policies, is to write them in simple English. An example of such a policy is the Simcoe and Erie General Insurance Company policy included in Appendix IV.

In general, there are five areas of risk which are not covered by professional liability insurance. These are business risks, unreasonably assumed risks, areas of specialized risk, activities outside the professional expertise of architects and engineers, and risks covered by other classes of insurance.

Risks which are inherent in doing business are generally uninsurable. If your client fails to pay your fees, the professional liability insurance will not cover your loss. If you contract to do an investigation for a fixed fee and lose money on the job, that is a business risk. If you find you are losing money on a job and you terminate your contract, your client may sue for breach of contract. This is again a business risk.

Risks which are inherent in doing business are generally uninsurable.

It is unreasonable for you to assume a risk which legitimately belongs to someone else and professional liability insurance will not protect you if you do and suffer a loss. Many contracts written by clients impose unreasonable risks on you. For example the consultant may be asked to agree to hold the client harmless for all injuries or damage to persons or property arising out of the project. Clearly it is unreasonable to expect you to assume the risks for activities over which you have no control. If you sign such a contract you may be liable for damages but your professional liability insurance will not protect you.

Hold harmless clauses are discussed in Chapter 8 on Contracts. The reader should review it to see the type of wording which may impose uninsurable responsibilities on the engineer.

Joint ventures are special cases of the assumption of liability of others. In a joint venture you are jointly and severally responsible with the other firms for the engineering services provided. This is somewhat similar to a business partnership where you are responsible for the entire debt of the partnership, not just your portion. In a joint venture you assume the professional liability of your joint venture partners, an unreasonable risk from the point of view of your insurer. It is possible to obtain a special insurance endorsement to cover a joint venture. The insurer will assess the claim record and experience of your partners in the joint venture and will write an endorsement or a separate policy to cover the professional activities of the joint venture.

The standard professional liability policy covers the general field of engineering. A policy for an engineering firm working in a special field such as geotechnical engineering must often be endorsed for that specialty and the insurer will satisfy himself that the firm has the expertise necessary to provide this service. The engineer should check as to what specialized fields he is insured for and avoid undertaking assignments in those fields for which he is not qualified in terms of his insurance policy.

Professional liability coverage is also virtually unavailable for certain hazardous waste disposal projects, specifically those associated with nuclear energy.

The professional liability insurance policy protects the engineer in his activities based on his professional knowledge. It does not protect him in activities outside his professional expertise. Giving advice on the economic feasibility of a project would be considered outside the field of the professional knowledge of an engineer or architect even though he may have experience in providing investment advice.

Professional liability does not cover risks which are covered by other classes of insurance. Clearly professional liability insurance cannot be expected to provide protection for the risks of operating an automobile or protection against fire or theft at the engineer's place of business.

The geotechnical businessman is then faced with the problem of selecting two independent variables (coverage and deductible) which control a third variable (premium) all three of which have significant influence on the continuity and profitability of his practice.

The professional liability policy may require that the insured consent to any settlement reached with the claimant. If the engineer does not accept a reasonable firm settlement offer which is acceptable to the insurer, and insists on carrying a dispute through the courts against the advice of his insurer, the insurer will not be obligated to pay a court Judgement beyond the amount he would have had to pay in the settlement.

As mentioned earlier, the cost of professional liability insurance will be a significant item in the overhead of most geotechnical firms. The policy is renewed annually and the premium is determined and paid at the time of renewal. The premium is usually based on four factors.

(1) The size and character of the coverage. This is usually the maximum annual claim ($250,000 to $2,000,000) for a given number of separate claims.

(2) The amount of the deductible payable by the engineering firm on each claim.

(3) The estimated volume of fees expected by the engineering firm for the coming year.

(4) The claim experience of the insurer for the category of the engineering firm and of the firm itself. Unlike automobile insurance, the claim experience of the engineering firm is not considered in setting the rate, only the claim experience of geotechnical firms considered as a group.

A rate based on factors 1, 2 and 4 above is determined and multiplied by the estimated fees to arrive at the annual premium for the professional liability insurance. Item 3 is a function of the economic times, and item 4 is outside the control of the geotechnical firm, although it is essential that in both cases the figures be correct.

The other two items, the total coverage and the size of the deductible are choices to be made by the geotechnical/businessman and they affect the amount of the annual premium. Increasing the coverage increases the premium while increasing the deductible decreases the premium.

The geotechnical businessman is then faced with the problem of selecting two independent variables (coverage and deductible) which control a third variable (premium) all three of which have significant influence on the continuity and profitability of his practice. He must ultimately balance the effect of the premium on his profits against the risk of paying out part or all of his deductible on a claim and the risk of a catastrophic loss which might exceed his coverage and bankrupt his business. There is no simple formula to arrive at this balance and the geotechnical businessman must arrive at a decision based on some assessment of his clients, his practice, his projects and his willingness to accept risk.

21
Case History X

A geotechnical firm was retained to inspect compaction in the reconstruction of a rural road. At one location re-alignment involved placement of a gravel fill about 1.5 m average thickness although locally the depth was up to 3 m. Compaction was carried out with heavy vibratory rollers.

Within a few days of completion of the fill a large landslide destroyed a section of the road and an adjacent rail line. The slide was 300 m long, 150 m maximum width and exposed a soft extra sensitive marine clay, on which planar lateral sliding took place at shallow depth.

A check of earthquake and rainfall records revealed no unusual events prior to the slide. Airphoto examination showed that no slides had occurred at this site within historic time. The natural landscape was relatively flat. However, geological information for the region indicated the probable presence of sensitive marine clay. The weight of the road fill and vibratory compaction were prime suspects for the triggering mechanism.

In this instance no claim was made against the geotechnical firm. If a claim had been made the courts might or might not have accepted a defence that the geotechnical firm was only responsible for measuring the density of the fill.

The geotechnical engineer was not asked to assess the risk of a quick-clay slide but if he had been asked, could he have delivered a satisfactory answer? Quick-clay slides have been studied for many years and vast amounts of research have been expended on back analyses of slides and on the mobilized shear strength properties involved on the sliding surfaces. Even if he had enjoyed an abnormally large budget it is doubtful if the consultant could have predicted the occurrence of this type of unusual shallow planar slide in relatively flat (1:10) ground. It would also certainly have been impossible for him to predict on any rational basis the extent of the actual land mass involved in the movement.

The behaviour of extra sensitive clay is only one of many areas in which the practicing consultant is working at the boundary of geotechnical knowledge and understanding. Others are seismic effects on foundations and slope stability, migration of toxic wastes through natural or artificial soil barriers, and risk assessments of rock falls and debris torrents. In these and other fields the geotechnical engineer may be giving advice and opinions on an incomplete understanding of the processes involved. Caution is the watchword.

22 Ethics

It is frequently difficult to distinguish between incompetence and dishonesty (Nasmith's Corollary)

Professional Engineering Associations include in their bylaws a section which outlines a code of ethics intended to govern the professional behaviour of their members. The code of ethics of the British Columbia Association of Professional Engineers included in Appendix III is typical of these codes.

"The professional engineer shall act at all times with fairness, loyalty and courtesy to his associates, employers, employees, and clients and with fidelity to the public needs. He shall approach his work with devotion to high ideals, personal honour and integrity."

This introductory paragraph is a general statement of ethical principles and the sections which follow specifically relate various aspects of the practice of engineering to these generalized principles.

Most business is conducted on the basis of mutual trust and an engineer who no longer trusts his client will find it well worth his while to extricate himself from the relationship as soon as possible.

A geotechnical engineer who has been robbed of his legitimate fees by a developer, who was underfinanced to begin with and avoided paying his bills by shuffling assets and liabilities between shadow companies, may wonder what professional ethics has to do with the real world of running a consulting business.

Nevertheless, ethical professional behaviour offers real benefits to the geotechnical engineer in the field of loss prevention in errors and omissions claims.

An engineer who consciously follows his own ethical standards should be able to recognize unethical behaviour in others.

Being ethical is not equivalent to being naive. An ethical engineer can and should recognize that his clients and others may not reciprocate his ethical behaviour and he must conduct his affairs so as to minimize the risk to himself of questionable practices on the part of his client. **Most business is conducted on the basis of**

mutual trust and an engineer who no longer trusts his client will find it well worth his while to extricate himself from the relationship as soon as possible.

Concern for and interest in the welfare of the client is one aspect of ethical behaviour which will pay dividends for the geotechnical engineer in a loss prevention program.

Geotechnical engineering is a very specialized field and few clients, whether they are laymen or engineers, have more than a very superficial understanding of the principles involved. Most in fact have no understanding. It then becomes the responsibility of the ethical geotechnical engineer to ensure that the client understands the interaction between his project and the art of geotechnical engineering. Since the client cannot be expected to learn much about geotechnical engineering it becomes the responsibility of the ethical engineer to find out as much as he needs to know about the client's project.

Ethical behaviour is learned more from example than from instruction.

It is not uncommon to hear a claim-threatened geotechnical engineer say "He never told me that the building was sensitive to settlement. All that he asked me for was a bearing capacity."

A client may not follow the geotechnical engineer's advice for a variety of reasons, some of which are quite legitimate. The ethical geotechnical engineer will ensure, that as far as possible, his client understands and accepts the consequences of rejecting the engineer's advice. The consequences must be described in written correspondence from the engineer to the client setting out in layman's language the engineer's opinion of the client's decision. This needs to be done even when the geotechnical engineer understands and accepts the client's reasons. This is not only a matter of ethical concern for the client's interest, it may easily be a matter of self preservation for the geotechnical engineer if negative consequences, however remote, come to pass.

The engineer's special knowledge imposes an ethical and legal obligation on him in relation to members of the general public. This may conflict with his relationship with his client. A developer may escape his obligation to the ultimate owner for poor workmanship or short-cutting on design leaving the engineer as a "deep pocket" against which claims can be made. If the geotechnical engineer can or should be able to foresee damage or risk of damage to some member of the public (e.g., an adjoining property owner) he has both an ethical and a legal responsibility to warn his client. Since corrective action, will in most cases, cost his client money the engineer is faced with the risk of alienating his client. If he does nothing he will be accepting substantial financial risk in the future. If risk of injury or death is involved, the engineer should have no ethical hesitation in pressing his warning as far as necessary. A written warning to the appropriate Authority having jurisdiction over the project would be in order.

In his ethical responsibility to the public the geotechnical engineer will find it to his advantage, from a loss prevention point of view, to dissociate himself from a client whose ethical standards diverge significantly from the spirit of the professional code of ethics.

Ethical behaviour is learned more from example than from instruction. A young engineer newly graduated from university may have high ethical standards in his personal life yet have difficulty in translating these standards into actions and responsibilities in the business world. Senior engineers in a consulting firm should emphasize, in their supervision of junior engineers the ethical responsibilities of the engineer to his client and to the public at large.

Ethical behaviour, as a loss prevention activity, will not exist in isolation. An engineer cannot practice ethical behaviour toward his clients and toward the general public without being ethical in his relationship to employers, employees and fellow engineers.

Photograph A illustrates an acceptable technique for an excavation in sensitive clay provided there is an adequate reserve against base instability.

Photograph B illustrates a procedure which can cause instability problems.

Did the geotechnical engineering study consider the effects of construction techniques on the performance of the excavation?

B

A

23 Summary

"The answer to the Great Question - of Life, The Universe and Everything" said Deep Thought, the computer, "is forty-two."

"Forty-two" yelled Loonquawl. "Is that all you've got to show for seven and a half million years' work?"

The Hitch-Hikers Guide to the Galaxy - Douglas Adams

The following notes summarize actions which a geotechnical consultant should take as a routine loss prevention activity.

When a proposal is being prepared or a client requests that you start a job, prepare an outline of the work you expect to do together with a cost estimate and schedule. This should be an internal confidential document and should include an assessment of risk. If the risk appears excessive and the economic climate permits it, you may wish to decline the assignment. Remember, it is much easier to turn down a high risk project than it is to escape from it once it is underway.

Know who your client is - have a contract with the client in writing and ensure that it clearly states what you expect to do and what you intend to provide to the client.

As the project proceeds ensure that any changes in your responsibility that develop are documented in writing. A carefully drawn agreement for test drilling and a foundation report is not very helpful in determining what the client expected when he is claiming for inadequate compaction control during construction.

Within the geotechnical firm establish an effective system for checking and review. The review system is particularly important to ensure that inexperienced engineers and new employees are trained to the standards of the geotechnical firm. The procedures should include a proper system of forms for recording data, as well as files for field observations, laboratory data, calculations, drawings and correspondence.

Establish a point of contact and line of communication with your client and maintain that line of communication. Be alert to any development of a personality conflict between your staff and the client's staff or the client's contractor. Record telephone conversations on a record-of-call form.

Many young engineers feel that they are hired to make decisions for the client rather than advise him.

Ensure that all risk-taking decisions are referred to the client. The risks of a project legitimately belong to the owner along with the profits. Fees charged by the geotechnical engineer are too small to permit him to accept any of the client's risk. You have a responsibility to inform the client of the uncertainties he is facing but he has the responsibility to make the decision. This is true even when the risk appears to be very small and the decision is one that the engineer agrees with and would make himself if he were the owner. It is still the owner's decision. This is a particularly important aspect of the internal review. **Many young engineers feel that they are hired to make decisions for the client rather than advise him.**

Finally keep good records and files. Memos regarding telephone conversations, field visits, meetings, etc., are vital in confirming what was done. When a job is completed, organize the file so that it presents a clear accurate account of activities. Cryptic undated notes on scraps of paper and masses of unidentified photographs, calculations and sketches may give a misleading account of the project in court and in any event can be used by an opposing lawyer to create an air of uncertainty, if not incompetence. Remember that when the job is completed you are simply tidying up the file but if a claim has been launched you may be "destroying evidence".

Once you are threatened with a claim for professional errors there are a number of actions you should take to minimize the risk of loss.

It is essential to remember the advice printed in large friendly letters on the cover of The Hitchhikers Guide to the Galaxy: - DON'T PANIC.

The first and one of the most important things to do is to maintain contact with your client. The worst thing to do is to ignore the problem in the hope that it will solve itself. If the problem has a geotechnical basis you will certainly be involved, and even if the connection to your work is tenuous or non existent, you may be involved in substantial cost in defending yourself against an unsubstantiated or frivolous claim.

If your client suffers a loss there is a very high probability that you will participate in that loss.

A principle to keep in mind is that **if your client suffers a loss there is a very high probability that you will participate in that loss.** The least damaging loss you will suffer will be unpaid fees while the most damaging will be a successful claim for professional negligence with attendant publicity and damage to your reputation. This principle should be kept in mind when you accept an assignment from a client. If he is likely to suffer a loss you will probably share in it.

If the problem is in any way a geotechnical problem your first concern should be to get the problem solved even if this means that you undertake work without any assurance that you will be paid. Staying on the job and solving the problem has at least three advantages.

1. - it will ensure that the problem is solved as quickly and economically as possible. This will minimize the loss which you may ultimately have to bear if you are found responsible for it.

2. - it will enable you to maintain your relationship with the client. The worst thing that can happen is for him to lose confidence in you and employ another geotechnical engineer. No matter how ethical and fair he is, another geotechnical engineer with the benefit of hindsight will find things that he would have done differently and in the mind of the client and his lawyer "different" will mean

"better". Unfortunately there is always a risk that the client will employ another geotechnical engineer who is neither ethical nor fair.

3. - finally in solving the problem you will know more about the technical and administrative aspects and how it developed than any outsider and will be in the strongest position to resist the claims of hostile expert witnesses.

An added bonus is that by maintaining your contact with the client and solving the problem you may actually be paid for your efforts.

When the threat of a claim arises, under no circumstances should you accept responsibility for the loss. No matter how damaging the evidence or how blatant your apparent error, it is almost certain that someone else, client, owner, contractor or other consultant had responsibility for some part of the error or aggravated the magnitude of the loss. If it is a continuing problem you should endeavour to be involved in finding a solution to the problem or investigating the causes. By the time the whole story unfolds it is unlikely that the geotechnical engineer is solely responsible for the loss and in many cases it will be found that his error makes a relatively minor contribution to the overall problem.

As soon as the possibility of a claim is recognized it is your responsibility as a part of your insurance coverage to notify your insurer. The insurer will normally appoint a legal firm or other specialist to evaluate the situation and monitor developments. You are responsible for keeping the insurer or his representative advised of your actions, and it is very useful to have any previous correspondence reviewed for its legal implications. Under no circumstances should you discuss responsibility for the claim without the input of the insurer or his representative.

At the same time it is advisable to open a file on the claim separate from the job file which deals with the project from which the claim originates. This file will be a confidential file not subject to seizure for evidence in court. The purpose of the file is to advise solicitors as the case develops. The lawyer will advise on the proper procedure for setting up such a file. This file will focus attention on developing a defensive position relative to the threatened claim.

If the claim continues to the stage where legal action is begun or threatened, the insurer will appoint a legal firm to act in the case. This stage does not usually occur until the project is largely complete and by then the geotechnical engineer is required to co-operate with the lawyers who will review the possibility of a negotiated settlement and prepare to file court documents.

Tales From Birdland

In Birdland things are organized in the usual way. Some birds are rich and some are poor.

Some go south for the winter and some stay home and watch TV programs from the south.

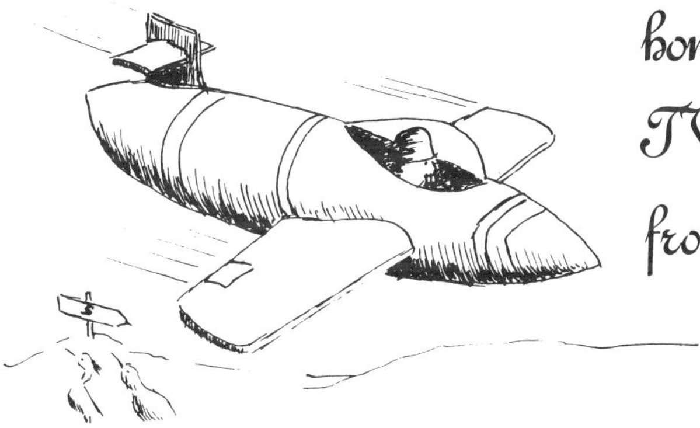

When communications break down disputes arise. Crows and ravens argue before an owl. After many moons judgement is given.

Storks and herons design buildings and are assisted by sandpipers and plovers who understand bending

moments and forces and that sort of thing.

Sometimes sandpipers designed buildings without consulting the herons and disputes arose. It was agreed that storks and herons would design

beautiful buildings and the sandpipers would design useful buildings.

The buildings are built by beavers. Beavers also build dams and bridges and recreation centres.

The herons and sandpipers were disdainful of the beavers because they were wealthy and used private aircraft instead of wings to go south.

In the years BT* many of the buildings cracked, some of the dams leaked and a few of the bridges sank into quicksand.

The beavers, the herons and the sandpipers agreed that these were "Acts of God".

A great prophet came from over the sea saying "God is dead, effective stress has established stability".

His followers were the gophers, a few at first then more and more until there was a great multitude.

The gophers were an inconspicuous lot, given to poking their noses in the dirt and muttering, "Two kips per square foot and a quarter of an inch settlement". Younger gophers muttered kilopascals, square meters and millimeters.

*Before Terzaghi.

The herons, sandpipers and beavers were happy because they could build bigger bridges and larger dams but they were unhappy because the gophers wanted more than chicken feed.

Still some of the bigger buildings cracked, and some of the larger dams leaked and some of the longer bridges sank into the quicksand.

The moral of this is that technology enables you to make bigger mistakes.

A Chilling Tale

The birds appointed a committee to plan and build a winter sports arena.

The chairman was a goose who had taken early retirement. His aunt owned a suitable piece of land down by the swamp. The secretary had a cousin who was a heron and could design the arena cheaply. One member was a retired beaver who knew about construction costs.

The chairman drew up a contract with the heron and got pledges from the partridges and a loan.

The heron said "We must employ a gopher to test the soil."

The gopher reported that the soil was wet and

silty and must be drained or frost heaving would occur.

The heron said "Indeed".

The gopher said "The installation of drains must be properly designed and inspected by an experienced gopher".

The heron said "Of course".

Chairman Goose said "The land cost more than we planned so we must economize".

The beaver said "I know where we can get a good supply of cheap drain rock".

The goose said "The heron must include the cost of all inspections in his fee."

Unfortunately the goose
suffered an accident
during a trip to the south
and the beaver had to
look after things for a while.

The winter sports arena was opened with
appropriate ceremony but presently the birds
found themselves curling uphill and downdale.
The winter bonspiel had
to be put off til spring.

The birds were unhappy. The banker was
unhappy. The new committee was unhappy.

The committee appointed a raven to sort things out.

The raven said to the gopher "I hope you have professsional insurance with the Sinful and Weary Company because someone is going to have to pay for this".

Question: How large was the gopher's deductible?

Answer: It doesn't much matter. By the time he extricated himself from this fiasco the gopher had contributed three quarters of his deductible to a negotiated settlement and ten times as many hours with the ravens as he had spent studying the site in the first place.

A Tale of Gratuitous Advice

The foxes and the beavers joined to build a condominium complex for the martins and hired a group of sandpipers to design and supervise the project.

When the project was complete the units were sold to families of martins.

When the paving in the parking lot started to break up the martins formed a committee to look at it. The committee asked a company of crows to enter a claim against the foxes, the beavers and sandpipers.

A group of ravens defended the sandpipers and reviewed all the construction records kept by the sandpipers.

The sandpipers didn't know much about soil but they knew about keeping records. The records

showed that while the beavers were paving the parking lot a firm of gophers had been asked to visit the site.

The senior gopher remembered that a junior gopher had gone to the site but no job file had been opened and no invoice had been submitted. The junior gopher had left the firm.

Question 1: Will the ravens representing the sandpipers benefit by "third partying" the gophers?

Question 2: (to be answered before question 1)
What is "third partying"?

Answer 2: Third partying is a procedure whereby a defendant in a professional liability lawsuit (the sandpipers) names a third party (the gophers) as having some responsibility for the alleged loss and so becomes a co-defendant in the suit.

Answer 1: In order to defend themselves the gophers will have to investigate the failure and when a negotiated settlement is reached they will probably have to contribute a portion of their deductible rather than be left to defend the claim alone.

SUGGESTED READING

The following list of papers are suggested as supplementary reading. These papers by Terzaghi, Casagrande and Peck deal with the philosophy of geotechnical engineering and although they do not address directly the problem of claims for professional negligence, the case histories and discussions often illustrate situations with a potential for litigation.

Terzaghi, K. 1958. Consultants Clients and Contractors. Journal of the Boston Society of Civil Engineers, 45, pp. 1-15. Reprinted, 1968 from Theory to Practice in Soil Mechanics. John Wiley and Sons. New York, N.Y.

Terzaghi, K. 1959. Soil Mechanics in Action. Civil Engineering, ASCE 29:2 pp. 33-34.

Casagrande, A. 1960. Discussion of Requirements for the Practice of Applied Soil Mechanics. Proceedings First Panamerican Conference on Soil Mechanics and Foundation Engineering, Mexico, III, pp. 1029-1037.

Casagrande, A. 1965. Role of the "Calculated Risk" in Earthwork and Foundation Engineering. Proceedings ASCE Journal of the Soil Mechanics and Foundations Division, 91, SM4, pp. 1-40.

Peck, R.B. 1962. Art and Science in Subsurface Engineering. Geotechnique, XII, pp. 60-66.

Peck, R.B. 1977. Pitfalls of Over Conservatism in Geotechnical Engineering.* Civil Engineering, ASCE 47:2, pp. 62-66.

Peck, R.B. 1973. Influence of Nontechnical Factors on the Quality of Embankment Dams.* Embankment Dam Engineering: The Casagrande Volume. John Wiley and Sons, New York, N.Y. pp. 201-208.

Peck, R.B. 1980. Where has all the Judgement Gone?* Laurits Bjerrum Minnefordrag No. 5. Norges Geotekniske Institutt, Oslo. Reprinted 1980. Canadian Geotechnical Journal, 17 pp. 584-590. Reprinted 1980. Norwegian Geotechnical Institute, Publication 134, pp. 1-5.

Peck, R.B. 1983. Nature Ignores Specialties. Geospec in Geotechnical News, 1:1, pp. 12-15.

*Reprinted 1984. Judgement in Geotechnical Engineering, The Professional Legacy of Ralph B. Peck. John Wiley & Sons, New York, N.Y.

ASSOCIATION OF SOIL AND FOUNDATION ENGINEERS

Since 1969, the Association of Soil and Foundation Engineers (ASFE) have been and continue to be active in risk management and loss prevention matters for consulting geotechnical engineers. ASFE publish many excellent articles and hold workshops on loss prevention procedures. For membership information and list of publications contact, Executive Director ASFE, Suite 225 - 8811 Colesville Road, Silver Springs, Maryland, 20910, U.S.A.

Appendix 1

The following *loss control bulletins have been reprinted with the permission of the National Program Administrator Inc. and Simcoe & Erie General Insurance Company.*

LOSS CONTROL BULLETIN

PROFESSIONAL LIABILITY LOSS CONTROL PROGRAM

BULLETIN NO. 52
February, 1981

Prepared by
NATIONAL PROGRAM ADMINISTRATOR INC.
in co-operation with
SIMCOE & ERIE GENERAL INSURANCE COMPANY

PROBLEMS INHERENT TO SOIL SETTLEMENT INVESTIGATIONS

By:
Claude Y. Mercier, F.I.I.C.
Executive Vice-President
National Program Administrator Inc.

Most design consultants are aware of the fact that post-construction soil settlement problems constitute some of the most severe and costly claims reported under this Professional Liability Insurance Program. During the investigation of a recent serious claim, the Insurers and their legal advisers had the opportunity to work in close co-operation with Dr. Arthur Casagrande of Harvard University, an internationally renowned soils expert. In the course of his investigation, Dr. Casagrande sent us an excellent letter which attempts to describe the complexity of these soil problem investigations in layman's language. Dr. Casagrande's letter is particularly interesting in that it stresses the unnecessary risks taken by the soils experts whose opinions may be empirical. With his permission, we have reproduced an edited report for the benefit of all Canadian design consultants.

"The purpose of this letter is to explain the reasons why (1) geotechnical investigators frequently differ widely in their prediction of magnitude and rate of settlements, and (2) why an investigator may change substantially his prediction as he proceeds from a pilot investigation to a comprehensive investigation of the subsurface conditions, and/or when reliable settlement observations become available. In part the reasons are the complexity of the mechanisms which control the volume decrease (consolidation) of clay soils, and which result in great uncertainties when trying to predict the magnitude of settlements and the rate at which they will develop. The reasons also lie in part in the human response of the investigator to the question how to report a wide range of uncertainty to his client. These two aspects will be discussed under the following headings.

Mechanics of Consolidation of Clay Soils

I will explain only the most basic aspects of consolidation, and I hope in simple terms. When a saturated bath sponge is squeezed, water is forced from the pores. To cause water to flow through a pipe, or through the voids of the sponge, one must create a pressure difference in the water, i.e. a hydraulic gradient. Pressure applied externally on a saturated sponge is felt by the water inside the sponge as an increase in pressure, but water at the outside surface of the sponge is unconfined and can escape freely. This difference in pressure between the water inside and at the surface of the sponge produces outward flow through interconnected pores. When squeezing a saturated sponge with very fine pores, it requires more time to compress it than to compress a sponge with coarse pores because water encounters greater resistance to flow in fine pores. If all pores on the surface of a saturated sponge were sealed, to prevent water from escaping, the pressure applied externally to the sponge would be carried only by the water in the pores, whereas the sponge skeleton would not feel the applied load, and it would remain soft.

The pores of a clay deposit below the groundwater level are usually saturated with water. The solid skeleton of the clay is like a sponge, but with pores (voids) of microscopic size. When a load is applied on a clay layer that has not previously been loaded (normally consolidated clay), the load is first carried entirely by the water in the pores, while the skeleton carries none of the load. If the clay layer is overlain and underlain by layers of pervious soils, and if the horizontal dimensions of the loaded area are much greater than the thickness of the clay layer, then the principal drainage path for the water that is squeezed from the clay is vertically upward and downward, into the pervious overlying and underlying soil strata.

Frequently, a clay stratum contains thin sand or silt layers, i.e. internal drainage layers, which will speed up greatly the rate of consolidation. The existence of such internal "drainage boundaries" may not be disclosed, or may only be partially revealed by a subsurface exploration in which only very few or no undisturbed samples are taken. An exception are clays which are thinly stratified with relatively pervious silt and/or very fine sand layers.

Each pair of layers is called a "varve". The rate at which such a varve clay stratum will consolidate depends greatly on the permeability characteristics of the more pervious layers in the varves.

When a thick clay stratum is loaded by means of long and narrow strip footings, or by small footings, drainage of the clay develops not only vertically but also laterally, which speeds up the rate of consolidation. Furthermore, the induced stresses rapidly decrease with depth, requiring analysis of the stress distribution within the clay stratum.

Last but not least, a settlement analysis requires reasonably reliable knowledge of the coefficients of permeability and compressibility of the clay layers, and whether and to what degree each layer has already been preconsolidated under a pressure in excess of the in situ overburden pressure. For this purpose one needs carefully conducted laboratory consolidation tests on a representative number of high quality undisturbed samples.

Thus, the number of variables that must be investigated, and the possible sources of serious misjudgments, is disconcertingly great. Based on my more than five decades of experience with problems in applied soil mechanics, and particularly with the recovery of undisturbed samples and performance of laboratory soil tests, I have learned already long ago, and preached it to my students and to my colleagues on many consulting boards, that there is only one rational approach: The investigator must assume for each parameter that enters in the analysis, a realistic estimate of its possible range. Then the result will usually end up in form of a wide range; sometimes an embarrassingly wide range which the investigator may hesitate to present to his client. E.g., the magnitude of the settlements may have to be stated within a range of plus/minus 50%, i.e. a ratio of one to three, or perhaps an even greater range. The development of settlements with time is even more difficult to estimate on the basis of conventional subsurface investigations and routine laboratory tests. For example, one may not be able to estimate the time required to reach 50% of the primary consolidation any closer than, for example, between two and ten years. Finally, the development of secondary consolidation is impossible to estimate from laboratory tests. Fortunately, in most cases the very slow development of the secondary consolidation and its small magnitude as compared to the primary consolidation, render this uncertainty unimportant by comparison with the magnitude of the primary consolidation. In general, it is the excessive magnitude of differential settlements due to primary consolidation that damage a structure and, therefore, the differential settlements are of principal concern in foundation engineering.

Settlement Estimates for Project "X" - With Comments on Differences Between Geotechnical Consultants

During a recent meeting, one of the participants requested that I comment on the great differences between statements that independent geotechnical consultants have made concerning the settlements of Project "X". The principal reasons for the differences are (1) the complexity of the subsoil conditions at the site, and (2) the uncertainties concerning the many variables involved when predictions must be made on the basis of insufficient information. An investigator may even change his own conclusions quite radically when he proceeds from a preliminary investigation to a comprehensive investigation of a given site.

Although I intended not to include in this presentation any mathematical formulas, I should make one exception to this self-imposed restriction. The time required to reach a certain percentage of primary consolidation increases with the square of the distance between effective drainage boundaries within a clay stratum. For example, if 50% primary consolidation and the corresponding settlement are reached when the drainage boundaries within a clay stratum are one foot apart, it would require $10 \times 10 = 100$ times longer to reach the same degree of consolidation if the drainage boundaries were 10 ft apart. This demonstrates the enormous influence that minor geologic details, that are often unknown, have on the rate of settlement which a structure will experience when it is founded on compressible, fine-grained soils.

At Project "X", the upper compressible stratum consists of highly plastic clay with a high water content (i.e. a highly compressible soil) which is divided by sand layers (drainage layers) into several layers of unknown thicknesses; and the lower compressible stratum consists of varved clay and silt, i.e. a thinly stratified deposit with a great number of drainage layers which result in the quick development of primary consolidation. Therefore, the first part of the time-settlement curve is produced chiefly by rapid consolidation of the varved (lower) stratum; and this is followed by much slower and ill-defined primary consolidation of the upper clay stratum, and onto which is superposed secondary consolidation of the varved clay stratum. In addition, we know from the few laboratory consolidation tests on more or less undisturbed samples, that these compressible soil strata are somewhat preconsolidated. From the geology and topography of the site it seems likely that the preconsolidation was not produced by a temporary surcharge of soil which was later eroded, but by several periods of low lake levels when sedimentation was interrupted and a temporary surface was exposed to drying. Therefore, probably there exist in these clay strata large variations in the degree of preconsolidation. The investigation of this preconsolidation would require a great number of consolidation tests on excellent undisturbed samples, i.e., a time consuming and costly investigation. To assume on the basis of one or two consolidation tests that the preconsolidation indicated by the tests applies to an entire stratum is liable to result in grossly over- or underestimating the degree of preconsolidation.

What would I have estimated for the range of

critical differential settlements between the core area and the outer walls of Project "X" in 1974, on the basis of the very limited subsurface exploration that ABC Consultants had available in 1974? Considering the complexity of the soil conditions at this site, and certain features of the proposed foundation design, I may have estimated a range of differential settlements between the core area and the outer walls with a lower limit approximately equal to ABC's estimate, or somewhat greater, and an upper limit about three times greater. Thus, the range may well have stretched over both, ABC's and XYZ's original estimates which were contained in their first reports.

ABC and XYZ both stated in their reports that their estimates of settlements were guided by their acquaintance with the performance of structures in the area. But they did not include any information on this important basis for their judgment of the probable settlements of the Project "X" Building. When faced with such problems, I have always leaned heavily on any reasonably reliable settlement observations that I could find for buildings in the vicinity of the site. I would have made every effort to collect such information, if I had been involved in an investigation for a new building in that particular area. It is for this reason that I had requested information on the performance of other buildings in the vicinity.

Permit me to end this letter, which has become too long already, with the following statement with which one of my colleagues summarized the subsurface investigations at a dam site:

> 'One boring - the geological conditions are perfectly clear.
> Two borings - serious doubts have arisen.
> Three borings - utter confusion.' "

ADDITIONAL COMMENTS

We have often stated that professional liability claims which result from soil problems are disproportionately frequent and costly. Our analysis of many of the files indicates clearly that one simple additional precaution rarely taken by design consultants would mitigate the problem to a large extent. We recommend that whatever the type of project, once the excavation has been completed, the soil consultant should always be called back to the site before commencement of the structural phase. The soil consultant should be given the opportunity to verify that conditions at the bottom of the excavated site are as predicted and that no part of the excavation shows soil conditions which are substantially different than those predicted. One prominent architectural consultant has told us that it has always been his practice to call the soil consultant back to the site upon completion of the excavation. This architect is convinced that this additional precautionary measure has eliminated many problems which would inevitably have occurred had construction continued without this additional precaution. This recommendation makes eminent good sense. We recommend this measure as a standard procedure which should be religiously followed.

LOSS CONTROL BULLETIN

PROFESSIONAL LIABILITY
LOSS CONTROL PROGRAM

BULLETIN No. 65
May 1983

Prepared by
NATIONAL PROGRAM ADMINISTRATOR INC.
in cooperation with
SIMCOE & ERIE GENERAL INSURANCE COMPANY

GEOTECHNICAL ENGINEERS AND FOUNDATION INVESTIGATION AND INSPECTION CONTRACTS

by:
Bryan S. Shapiro, L.L.B.
BULL, HOUSSER & TUPPER

In the geotechnical area of engineering, a common practice which constitutes a danger to architects, engineers and owners has recently come to the attention of the design professions. Fortunately, it would appear that the fate of the professions is in their own hands in this particular matter.

Geotechnical engineers are often engaged directly by the owner for the initial foundation investigation. This, in fact, is the recommended practice because of the sensitive nature of geotechnical practice, and further because errors in this area tend to have rather significant financial consequences when relating same to the size of professional liability claims against the design professions. As indicated in earlier bulletins, if the architect/engineer engages the geotechnical engineer, then he becomes vicariously liable for the latter's errors, omissions and negligent acts in carrying out his geotechnical services.

Therefore, the standard agreement between Client and Engineer (A.C.E.C. Document No. 31) in Article 2.2 contemplates the owner engaging the services of the geotechnical consultant, with the prime consultant architect/engineer reasonably entitled to rely upon the accuracy and completeness of such information and data furnished by such consultant by or through the owner (Article 2.1).

It now appears that a practice has built up whereby the foundation inspection portion of the geotechnical services are, by way of requirement of the architect/engineer on the project, being relegated into a separate contract. The geotechnical engineer is being required by the architect/engineer no longer to contract with the owner for this vital portion of the geotechnical services but is now directed to contract as a subcontractor to the contractor on the project. Furthermore, and even more dangerous, is the apparently misguided requirement of the architect/engineer that payment of the geotechnical consultant according to such subcontract arrangement be by way of cash allowance included in the contractor's general contract.

This is a most dangerous practice which inherently subjects the geotechnical engineer to a conflict of interest position at best and, in our opinion, seriously prejudices the position of the owner on the project.

In no instance should the party (the contractor) charged with the responsibility of doing the work have control over its inspection. In the situation described above, where the contractor is required to engage the geotechnical consultant, the contractor actually has control over the final report which is in fact forwarded to the owner through the latter's agent, the architect/engineer.

Many times in the past comments have been expressed about the importance and critical nature of the field services rendered by architects and engineers whereby the design professions are required to ascertain that the work of the contractor generally conforms with the contract documents prepared for the project.

In the sensitive area of geotechnical engineering and construction related thereto, it is all the more critical that the geotechnical engineer has an opportunity to actually inspect the subject matter of his work and to do so in a manner which is solely within the purview of his own discretion for the reasons expressed in the *Guidelines to Engineering Practice* which have been distributed by the National Program Administrator through the A.C.E.C.

To tie the geotechnical consultant's professional discretion for rendering field services to a lump sum cash allowance included in the general contract is dangerous from the owner's point of view for the reason expressed above as it gives the contractor control over the final dissemination of the geotechnical report. The geotechnical consultant might find himself in the classic *Surrey v. Carroll Hatch & Associates Ltd.* situation.

In that particular case, which has been described in earlier bulletins, the reader will recall that the British Columbia courts appear to take the position that the subconsultant engineer almost had a duty of care to go over the head of his prime consultant to warn the owner on the project where the prime consultant was not apparently following the subconsultant engineer's opinion as to the necessity for further soil analyses. This is the type of situation that arises in cases wherein the contractor engages the soils consultant.

In our opinion, the owner should engage the soils engineer for both the foundation investigation as well as the foundation inspection. To do otherwise is to require a useless dichotomy of service which can only operate to the prejudice of the owner. Architects and engineers who are preparing specifications requiring such a dichotomy are not only doing their clients a disservice but are subjecting themselves to future professional liability claims as well.

Therefore, the following recommendations are made with regard to the services of geotechnical consultants on projects:

1. The owner should engage the geotechnical engineer pursuant to a single contract for all aspects of the foundation investigation and foundation inspection.
2. In no instance should the architect/engineer specify that the foundation inspection is to be carried out pursuant to a separate contract between the geotechnical consultant and the contractor.
3. Whenever the geotechnical consultant has not been hired by the owner to provide inspection services, then the geotechnical engineer's contract with the owner should clearly indicate (and their report on the foundation investigation should also specify) that they are not a party to the construction process and will not be reviewing same on behalf of the owner or anyone else. Therefore, as a result, they will be unable to confirm that actual site conditions as expected from the geotechnical investigation may not pertain and they will therefore not be responsible for any situations which may arise on the project subsequent to the completion of their services in connection with the foundation investigation.
4. The architect or engineer should never take it upon himself to remit only part of the soils report he has prepared to bidders or contractors. The entire report should be made available as any decision of the architect or engineer to keep part of the information back may well result in later claims alleging that the consultant, as a professional, negligently misinformed the contractor.
5. The consultant should always, in his specifications and on any drawing where bore-holes are shown, incorporate the following clause:

   "Any information pertaining to soils and all bore-hole logs are furnished by the architect or engineer as a matter of general information only and bore-hole descriptions or logs are not to be interpreted as descriptive of conditions at locations other than those described by the bore-holes themselves."

LOSS CONTROL BULLETIN

PROFESSIONAL LIABILITY LOSS CONTROL PROGRAM

BULLETIN No. 66
July, 1983

Prepared by
NATIONAL PROGRAM ADMINISTRATOR INC.
in cooperation with
SIMCOE & ERIE GENERAL INSURANCE COMPANY

CERTIFICATES OF SUPERVISION AND CONSTRUCTION - DANGER FOR THE ENGINEER

by:
Bryan S. Shapiro, L.L.B.
BULL, HOUSSER & TUPPER

A recent survey was commissioned on behalf of the A.P.E.B.C. Consulting Practice Committee of lower mainland cities' and municipalities' current practices in relation to Certificates of Supervision and Construction. As most members are undoubtedly aware, these Certificates are drafted by the individual municipalities and cities and are therefore quite different in their wording. There appears to be no standard or common phraseology between the various forms of Certificates which, when studied closely, disclose themselves as constituting a dangerous trap for the unwary engineer.

The intent of these Certificates is apparently twofold. Firstly, the municipalities in an effort to protect their own field inspectors, are attempting to ensure that responsibility for the failure to detect contractors' deviations from the design does not affect the liability exposure of the municipalities and their inspection staff. Secondly, and perhaps just as important to the municipalities, there is the altruistic principle that the public interest must be served insofar as health, welfare and safety is concerned, and the requirement for the Certificates of Supervision and Construction would seem, on the surface, to be a safeguard in this area.

The philosophical conflict which arises in all of this is the contractual mandate which exists between engineers and their clients. It may or may not be the case, on any given project, that an engineer has obtained contractual terms of reference from his client, which will allow him to render a level of supervision of the construction work, which will allow him to indeed put his name and seal to the form of Certificate required by a particular municipality. Keep in mind that these Certificates will ultimately constitute certifications or representations on the part of the engineers executing these documents which will be relied upon both by the municipalities concerned and by the engineer's own clients. There is a potential detriment to these parties in the event that the certifications are given in error. This can result in enormous professional liability implications for engineers executing such documents.

The overriding principle to consider is that no engineer can certify that which he has not seen. Similarly, no engineer should be responsible for the contractor's means, methods, procedures, sequences, techniques of construction or safety measures with respect thereto. We then look at the logistical problem created when an engineer purports to execute a Certificate of Supervision and Construction when he does not have a contractual mandate to perform a level of supervision which will allow him to issue such a certification without qualification. An engineer cannot certify in a vacuum. In other words, to certify sight unseen could be characterized as fraud, or at best, professional negligence. In either case, the ethical problems as well as professional liability implications for the engineer are legion.

The definition of "Field Services" contained in Article 1.8 in the latest draft of the Association of Consulting Engineers of Canada Standard Agreement between Client and Engineer indicates that the only field services (supervision) which should be rendered by professional engineers is the sole professional discretion field service mandate, which in effect allows the engineer to be on the site selectively at his discretion. Such factors as the reliability of the contractor and the sensitivity of the design and construction will ultimately affect the engineer's discretion in this regard. Therefore, the individual requirements of the Certificates of Supervision and Construction may bear no relation to the actual terms of reference (i.e. his contract) for the individual engineer on a project. Therein lies the danger and the source of most professional liability claims against professional engineers. Engineers are certifying without in fact having first hand knowledge of the subject matter of their own Certificates.

With the above scenario in mind, let us now look at some typical wordings of Certificates of Supervision and Construction from various municipalities and cities in British Columbia.The italics in the following examples is provided in order to emphasize particular areas of danger for the engineer.

Example No. 1:

> "The owner shall certify that he has engaged a professional engineer experienced in the field of municipal engineering to carry out survey, design, *field inspection* and preparation and certification of *as-built drawings*.Field inspection shall consist of *general and sufficient resident inspection* to *ensure* that the works and services *are constructed* and installed *in accordance with* the plans as approved for construction, and the standards as herein contained. Sufficient *resident inspection* shall range from a *minimum* of one site visit per day during the active construction period to *full time*

resident inspection for major developments as determined by the municipal engineer. The professional engineer shall acknowledge that he has been engaged in this capacity."

This example illustrates the most common error contained in the Certificates of Supervision and Construction throughout the various municipal documentation. It indicates the typical error of attempting to quantify a level of site presence by the engineer. It is this very quantification of the extent of field services to be offered which is the root cause of the majority of professional liability claims against engineering design professionals . Only the sole professional discretion field service mandate, as defined by the A.C.E.C. Standard Agreement, should be used. Furthermore, the word "ensure" implies a guarantee or warranty by the professional engineer that the works and services are indeed constructed and installed in accordance with the plans as approved for construction. Engineers who guarantee and warrant a result will not have the benefit of their professional liability insurance in the event that claims arise which arise out of such guarantees. It is virtually impossible for an engineer, at the inception date of his contract, to indicate a level of field service responsibility based upon a preconceived level of site responsibility because of the various parameters of the project (including the design) of which he can have no knowledge at the time.

Example No. 2:

"I hereby certify that the plan showing the works as *actually constructed* is correct and that to the best of my knowledge the works are complete and constructed *in accordance with* the specifications and standards contained in..."

No engineer can certify the contractor's as-built work as no engineer can possibly certify every aspect of the contractor's workmanship, whether or not the engineer is at the site 100% of the time. Furthermore, no engineer can certify that the work is complete and constructed "...in accordance with the specifications...", for the very same reason as expressed below.

Example No. 3:

"I hereby acknowledge that *I have been engaged* by the owner *to carry out* design, *field inspection* and preparation and certification of *as-built drawings* in accordance with..."

The above example is another indication of a common pitfall for professional engineers. To indicate in such a Certificate that he has a mandate to carry out field inspection amongst other things means nothing. As stated above, such so-called "field inspection" must be defined with reference to the actual contract existing between the engineer and his client. If possible, the actual written agreement between the engineer and his client pertaining to his field service responsibility should be annexed to the Certificate being executed by the engineer. If the client has in fact not engaged the engineer to provide a level of field inspection commensurate with the basic philosophy flowing from the A.C.E.C. definition of Field Services, then the engineer should never affix his name and seal to the form of Certificate referred to in this example, other than in a situation where the engineer also attaches a suitably drafted qualification which makes the Certificate virtually not worth the paper it is written on.

Example No. 4:

"Engineering inspection shall mean taking such steps within the *scope of my authority* reasonably required by good practice for the *execution* of the elements of the work *in accordance with* the plans, specifications, drawings and designs which are approved by the minicipality for a building permit."

The common error contained within this example again relates to the fact that the engineer's own contract with his client may not allow him to provide a level of engineering inspection commensurate with "good practice" as those words are used in the example Certificate. Even more dangerous, however, is the suggestion that the engineer has authority "... for the execution of work". The engineer does not "execute" the work, he only reviews or supervises same in order to ascertain "general" conformance of such work with the design represented by the contract documents. The word "execute" in this context should always be removed from any such certification documents. Also, the engineer can never certify that the work is more than in "general" accordance with the plans, specifications,etc.

Example No. 5:

"The engineer is responsible for the *engineering aspects of construction* ."

This example wording is very ambiguous and suggests that the engineer is again somehow responsible for the contractor and his construction methods, techniques, etc. This is intolerable from the engineer's point of view. He has no contractual obligation to the contractor and again is only remunerated pursuant to his contract with his client for ascertaining general compliance of the contractor's work to his design.

Example No. 6:

"The engineer, or *another suitably qualified person responsible to the engineer,*shall *review all shop drawings* and other related documents relevant to the design *to determine conformance with the design.*"

The use of the words "or another suitably qualified person responsible to the engineer..." in this example suggests that the engineer will be responsible to engage a subconsultant or employee of his own to review shop drawings for conformance with the design. This immeasurably increases the chances of claim against the engineer since rarely are engineers engaged and properly paid to review shop drawings in detail so as to be able to certify that they are,in all respects, in conformance with the design.

Suppliers, manufacturers and others may engage their own personnel or other engineers to spend hundreds or even thousands of hours preparing the detailed designs inherent in shop drawings submitted to the engineer for review on a typical construction project. At best, most engineers are only paid to give such drawings a cursory overview so as to ascertain that such drawings conform to the dimensional requirements of the project, and in order to determine that such drawings are also in general conformity with the overall design concept for the project.

If the engineer executes such a Certificate relating to shop drawings, he will be inheriting any errors or omissions contained within the shop drawings prepared by others, which errors can only be ascertained by spending the same quantum of hours spent by the original designer. Since the engineer is not paid for such a review, he must not sign such a Certificate unless he is indeed properly mandated and paid to conduct such a review or unless such a Certificate is properly qualified by the engineer.

Example No. 7:

> "This is to certify that I am the engineer responsible for supervising the construction on this project. Where necessary, in my opinion, I will engage specialist engineers experienced in these works as well as employing technicians or inspectors to *ensure acceptable methods,* plans and specifications *are followed.*"

This example again illustrates the common error of requiring a guarantee from the engineer in that he is to "ensure" acceptable methods, etc., which the engineer cannot do. At best, the wording should read"...employing technicians or inspectors to ascertain that the construction of the work generally conforms to the plans and specifications prepared for the project." Furthermore, the contractor's methods of construction are not the responsibility of the engineer and he should not insert himself into such a dangerous realm.

Example No. 8:

> "The engineer has been informed that while the engineering department of the city will pass on information it has concerning the location of plant and utilities, upon request, *it does not guarantee*, in any manner, the *accuracy* of such information and the *city shall not be liable* for any loss or damage *resulting from the inaccuracy* of such information, *whether such inaccuracy results from negligence* or otherwise howsoever."

In this example, the city is asking to be exhonorated from responsibility for faulty information which it provides to the engineer, whether such information results from negligence or otherwise. This is totally unacceptable from the engineer's point of view, and again, Article 2.1 of the A.C.E.C. Standard Form of Agreement should be referred to in this context. Article 2.1 indicates that the engineer is entitled to rely upon the accuracy and completeness of information and data furnished by or through the client (city), and where such information or data is in error, then the engineer shall not be responsible to the client (city) for the consequences of any error or omission contained therein. This is only reasonable, as the use of the wording set forth in the example itself would allow the city and its employees to provide sloppily prepared information to the engineer with impunity. This, in itself, is detrimental to the public interest and safety.

CONCLUSION:

Since there are problems associated with most, if not all, of the various forms of Certificates of Supervision and Construction used by cities and municipalities in British Columbia,it will be up to the engineer to be prudent and careful whenever he is asked to sign any of the commonly employed standard forms of Certificate to ascertain that the wordings used are acceptable and appropriate to the engineer's contractual mandate on a particular project. In the event that such is not the case then the engineer has only a few alternatives. Firstly, he can always decline to accept the engineering commission. However, in difficult economic times, this is a tough decision to make. Secondly, he can take on the commission and issue suitably qualified Certificates. Such qualifications should contain phraseology such as the following:

> "The engineer neither warrants nor represents the accuracy or completeness of this Certificate of ______ as the engineer has not had an opportunity to inspect the subject matter of this Certificate to the extent that he deems to be necessary in his professional opinion."

Anyone relying upon the aforementioned qualified Certificate will then do so at their own risk. The further danger to the engineer remains, however, that even if the city or municipality or client, as the case may be, cannot bring suit against the engineer for errors contained within such Certificates,third parties who are injured as a result of the engineer providing less than a "full" field service may still bring claims against him, and they will not find their claims barred or otherwise diminished by the aforementioned qualification to which they are not a party.

It is therefore submitted that where an engineer perceives a potential danger to life or property arising out of the provision of field services which represent less than he professionally feels are necessary in order to protect the public interest, then he may have an ethical problem in preparing an unqualified or even a qualified Certificate of Supervision and Construction for that matter. In view of the pitfalls and dangers lying in wait for unsuspecting engineers who cavalierly complete and affix their names and professional reputations to the various forms of Certificate documentation referred to in this article, engineers would do well to submit any Certificates of doubtful or ambiguous meaning to their legal counsel for review prior to executing or even agreeing to execute such documentation.

LOSS CONTROL BULLETIN

PROFESSIONAL LIABILITY
LOSS CONTROL PROGRAM

BULLETIN No. 68
December, 1983

Prepared by
NATIONAL PROGRAM ADMINISTRATOR INC.
in cooperation with
SIMCOE & ERIE GENERAL INSURANCE COMPANY

THE DESIGN CONSULTANT'S LIABILITY FOR FIELD SERVICES

Part I

by:
Claude Y. Mercier, F.I.I.C.
President
National Program Administrator Inc.

This is the first of two bulletins dealing with this important topic. Part I describes the design consultant's liability to the owner and suggests ways of reducing the chances of allegations of negligence in the rendering of field services. Part II will deal with the consultant's relationship with the contractor.

Architects and engineers consider themselves designers first and foremost, and project coordinators/managers second. A look at the makeup of their fee schedules will make this obvious. Usually, anywhere from 70 to 90% of the consultant's fee is applied against design services with the remaining relatively low fee being charged to construction or field services. Unfortunately, the scope of the consultant's professional liability is not proportional to the apportionment of fees (and resulting time and attention) between the design and the construction phases of the mandate respectively. In fact, during the last ten years, less than 50% of the professional liability claims advanced against Canadian design consultants were the result of a slip of the slide-rule. Relatively few claims can be definitely and exclusively classified as "design error". Well in excess of 50% of the claims contain a major element of allegations of negligence in the rendering of field services by the consultant.

Architects and engineers in private practice should recognize the great extent of their liability to the project owner and even to third parties (such as subsequent purchasers of the property or tenants) for their failure to detect bad workmanship or substitution of materials. Once the design consultant **and all his employees** understand the breadth and scope of their liability for field services, there are measures which can be taken to help reduce the likelihood of claims.

The purpose of this bulletin is certainly not to encourage consultants to "weasel" out of their proper responsibilities in an inequitable manner. Our purpose is to encourage architects and engineers to define and limit their liability simply by making it commensurate with the scope of their mandate from the owner and the amount of their fee. There is nothing equitable about being held liable for defects the consultants had absolutely no chance to detect during the occasional site visits for which they were being remunerated.

It is a matter of equity and professional ethics that design consultants should be answerable for errors and omissions in the discharge of their professional mandate. It should also follow that where the client or owner has deliberately tied the consultant's hands by restricting his field services mandate, then, that owner should hold the consultant harmless for bad materials or workmanship which could not possibly have been detected under such conditions.

How can architects and engineers reduce the risk of professional liability claims alleging negligence in the rendering of field services?

Educating the Client

Every design consultant has experienced the client who wishes to save money on an already strained construction budget. Experience has shown that one of the first elements of the construction package the owner will try to reduce or even excise is the mandate for field services to be rendered by the design professional. Out of all the possible avenues open to him to reduce costs, why would the client look at the consultant's field services mandate as the least painful cost-cutting opportunity? We suggest that this is due to a lack of understanding on the part of clients or owners of the importance of a proper field services mandate. Perhaps a great deal of the responsibility for this misconception on the part of clients can be laid at the feet of the design professions who have, for the most part, been remiss in properly explaining to their clients the dangers inherent in limiting the field service aspect of the overall professional mandate. This is an educational responsibility on the part of design consultants, who must sit down with their clients at the time of initial engagement, and explain the substance of the services to be offered throughout the various phases of the project, and emphasize the importance of each service rendered to the success of the overall project. The consultant should explain to the client that he cannot prudently accept responsibility for reviewing the execution of his design without an appropriate mandate to oversee its implementation.

Educating the client is simply a matter of helping him understand that the design consultant cannot be held liable for that which he has not had a chance to review. It is suggested here that Standard Document 31S of the Association of Consulting Engineers of Canada and the accompanying guidelines available from A.C.E.C. are ideal instruments to help the design consultant in this educational process. It is further suggested that, although these standard documents are promulgated through the Association of Consulting Engineers of Canada, they can also be used by architects in the same manner.

Finally, during this process of negotiations with the client, the design consultant should further point out that if his field service mandate is restricted, then any form of certification which the design consultant will have to prepare and sign later, will have to be qualified to reflect such restrictions.

Hopefully, once the client better understands the importance of proper field services, he will let the consultant exercise his professional judgement and allow him to provide the level of on-site presence which he, in his professional discretion, deems necessary under the specific project circumstances.

If the client still insists upon restricting the consultant's field services mandate for reasons of economy, then the consultant must accurately define his restricted mandate in his contract with the client and limit his liability accordingly.

The Owner/Consultant Agreement

Standard Contract No. 31S of the Association of Consulting Engineers of Canada offers three options for spelling out the engineer's field services mandate. The first option is for the rendering of full field services as deemed necessary by the engineer in his professional discretion. The second option is for the rendering of full field services as regards only specific phases of the projects which are listed in an addendum attached to the contract. In the case of the second option, the contract clearly restricts the consultant's liability to those phases of the projects for which he was allowed to render proper field services and to the exclusion of all other phases of the project. The third option is one wherein the engineer renders no field services whatsoever to the client with a clear written understanding that whenever the engineer is invited on site to deal with a specific problem, his liability will be restricted to the advice he has given in connection with that problem and nothing else.

Certification

The number of certificates of various forms which design consultants must sign has tended to increase in recent years. Progress certificates are required by the bonding company and by the project financiers; certificates will be required to confirm substantial performance; certificates will be required by the various authorities before occupancy permits can be issued. All these certificates have one thing in common: they require the design consultant to attest that the project was built according to plans and specifications. The recipients of these certificates unquestionably rely upon them to the extent that they are in a position to suffer great financial loss, should they be in error. The same large question looms. How can the design consultant certify that all phases of the project have been built according to plans and specifications if his field services mandate did not allow him to verify that this is indeed the case? In other words, how can the design consultant be expected to certify that which he has not seen?

It would not be logical for a client to expect a design consultant to certify the proper construction or erection of something he has not had a chance to see. If the installation of the insulation by the contractor was not one of the phases of the project listed on the addendum attached to the contract, then the client cannot expect the design consultant to certify proper installation of the insulation and to sign any form of certificate attesting that the building meets energy conservation standards. Either the design consultant is paid to review the installation of the insulation which will allow him to certify, or he is not paid to do so and he cannot certify that particular part of the project.

Obviously, the design consultant can issue a completely unqualified certificate only if he has been allowed to render the extent of field services which he, in his professional discretion, deemed necessary under the project circumstances, when the mandate was first negotiated with the owner. If, at that time, the owner insisted on a restricted field services mandate, then the consultant should have warned the owner that all certificates would be qualified whether this is acceptable or not by the recipients of these certificates. At the time of issuing the certificates, the consultant has no choice but to add a paragraph on the actual form or attach an addendum to it qualifying the certification to reflect the extent of the field services mandate. If the field services mandate has been properly defined in the contract between the consultant and the owner, then the qualifying words on the certificate should be as close as possible to the contractual description of the field services mandate.

Our courts have held that professionals are liable towards anyone they knew, or should have known, would rely on their professional opinion. Is it fair then for consultants to be held liable to pure strangers such as banks, bonding companies, and municipal authorities for statements made in an unqualified certificate which attests to the acceptance of work which they have not had a chance to see and which they were not remunerated to review?

The guidelines accompanying A.C.E.C. Standard Document 31S contain suggested phraseology for the qualification of certificates.

Conclusion

An autopsy of a large number of our claim files has shown that our courts tend to hold design consultants liable for improper materials or workmanship on the part of the contractor following a failure of the design consultant to detect the discrepancies during the construction process. It is also clearly evident that design consultants will be held liable for the statements contained within the certificates they issue attesting that the project was built according to plans and specifications.

In light of these trends, it is necessary for design consultants to render a full field services mandate as they deem necessary for a given project and to be remunerated in accordance with the extent of that mandate. If, for reasons of economy, that mandate must be restricted, then the consultant's liability should be restricted and all certificates should be qualified accordingly.

The implementation of these recommendations starts with a proper education of the client at the very beginning of negotiations before the project ever gets under way. Finally, it should be noted that under certain provincial Building Codes, it is mandatory that implementation of the design be done under the supervision of a professional engineer. The Codes do not seem to indicate that these field services must be rendered by the original designer, but simply that an engineer must be on site. It is suggested that where the owner will not allow his design consultant to render the level of professional services required, then the owner should be put on notice in writing that he may be in breach of the law.

LOSS CONTROL BULLETIN

PROFESSIONAL LIABILITY
LOSS CONTROL PROGRAM

BULLETIN No. 69
February, 1984

Prepared by
NATIONAL PROGRAM ADMINISTRATOR INC.
in cooperation with
SIMCOE & ERIE GENERAL INSURANCE COMPANY

THE DESIGN CONSULTANT'S LIABILITY FOR FIELD SERVICES

Part II

by:
Claude Y. Mercier, F.I.I.C.
President
National Program Administrator Inc.

Loss Control Bulletin No.68 dealt with the onerous burden of liability towards the owner which falls upon the shoulders of the design consultant and which emanates from the construction phase of the consultant's mandate. The employees delegated by the architectural or engineering firm to the construction site to provide the field review services defined in the consultant's contract should all be aware of their liability for failure to detect bad workmanship or materials. As stated in Loss Control Bulletin No. 68, that liability should be defined as accurately as possible in the contract document between the consulting firm and the owner.

The consultant's or his representative's presence in the field will lead to close interaction with the contractor. Usually, there is no contractual relationship whatsoever between the architect or engineer and the contractor. In the so-called "triangular relationship", the consultant's contract is between himself and the owner and similarly the contractor's contract is with the owner. There are no contractual arrangements between the consultant and the contractor. Inevitably, however, in the process of rendering field services, the consultant will be in constant contact with the contractor and with many of the subcontractors or trade contractors. It has been explained in previous bulletins that professionals are liable towards all those whom they knew, or should have known, would rely on their professional advice. Contractors certainly fall within that category of individuals or organizations whom consultants are aware are relying upon their professional advice. In their dealings with contractors during the construction process, architects and engineers should therefore recognize their exposure to allegations of negligence in the nature of the professional advice they have given.

How can the design consultant provide the level of field services he has defined in his contract with the owner, while doing everything possible to reduce the chances of claims emanating from the contractor?

CONSTRUCTION METHODS

The constultant's representatives or employees on site should be aware of the fact that the dual role of the design consultant is first, to design the project for the owner and, second, to review the construction process to determine that the project is being built in general conformity with the design. It is the contractor who is responsible for the ultimate delivery of a project which is in conformity with the design. It is the contractor who determines how he will go about it.

Contractors themselves recognize that they are responsible for construction methods. In fact, Standard Construction Document No. CCDC2 Article GC3.3 reads as follows: "The Consultant will not be responsible for and will not have control or charge of construction means, methods, techniques, sequences or procedures, or for safety precautions and programs required for the Work in accordance with the applicable construction safety legislation, other regulations or general construction practice. The Consultant will not be responsible for or have control or charge over the acts or omissions of the Contractor, his Subcontractors or their agents, employees or other persons performing any of the Work".

CCDC Document No. 2 was prepared by the Canadian Construction Documents Committee, which is a joint committee composed of representatives appointed by various associations of the construction industry and including the Canadian Construction Association. Therefore, the national association representing contractors has participated in the preparation of CCDC Document No. 2, wherein the contractor, in his contract with the owner, recognizes that he is the one responsible for construction means, methods, techniques, sequences or procedures.

Many times during the construction process, the contractor will seek the advice of the consultant's field representatives, usually to obtain clarifications or explanations as to the design intent. The consultant's field representatives should remember that they are not responsible for construction methods and their advice to contractors should therefore be very carefully worded to ensure that they help the contractor in determining what he should do, but not how he should do it.

All meetings and discussions between the consultant and the contractor and all decisions emanating therefrom should be documented in writing and these records should be kept indefinitely. The consultant's field representatives should keep a diary wherein they log daily notes as to what is going on on the construction site and also they should make sure that someone is appointed to prepare minutes of all site meetings. These minutes must then be distributed to all meeting participants. It should be made clear to the consultant's employees that these notes, minutes and diaries are the property of the consulting firm and not of the employees themselves, who should not take these documents with them if and when they ever leave the firm. These records must be kept by the consulting firm with the project file, as they could become key documents in the defence of possible future professional liability claims, sometimes many years after completion of the project.

CONSTRUCTION SITE SAFETY

Article GC3.3 of CCDC Document No. 2, which is quoted above, also states quite clearly that it is the contractor who is responsible for site safety and not the consultant. It is a good idea to incorporate a similar clause in the owner/consultant agreement firstly, to make sure that the owner/consultant and the owner/contractor agreements are compatible and secondly, to stress the fact that site safety is the responsibility of the contractor.

Contractual agreements making contractors responsible for site safety do not totally relieve consultants of their responsibility for the safety of construction workers and members of the public. As professionals, architects and engineers have a responsibility to the world at large for making sure that blatant violations of safety codes and procedures are not tolerated. The courts would not look kindly upon the design consultant who has ignored a dangerous situation simply because his contract with the owner states that it is the contractor who is responsible for site safety. When the consultant's field representatives notice violations of safety standards, they should immediately point the problem out to the contractor and ask him to correct the situation. In the absence of cooperation on the part of the contractor, the consultant should then call in the proper authorities who will, undoubtedly, either make sure that the contractor rectifies the problem or close down the site. Contract clauses which state that the contractor is responsible for site safety are useful in determining who is responsible for doing what. They are not a license for consultants to ignore obvious dangers to life and limb.

INFORMED CONSENT

Often during the construction process, the contractor will make suggestions to the consultant proposing certain design changes or substitutions of material with a view to providing an economy to the owner. Some of these suggestions may be perfectly acceptable to the consultant, who may feel that the alternate design or material will fit perfectly well the intended purpose while indeed reducing the owner's costs. In a spirit of cooperation with the contractor, the consultant may well approve the alternative.

In other cases, however, the consultant may feel that the suggested change is a compromise which may save the owner money immediatley but work to the owner's disadvantage in the long run. For example, the contractor might suggest an alternate roofing system which would reduce construction costs but which would ultimately increase the owner's maintenance costs. Under such circumstances, we suggest that it is not the consultant's responsibility to make the decision to either accept or reject the contractor's suggestion.

Our courts in Canada have upheld many times in recent years what is known as the "theory of informed consent". According to this theory, it is not the responsibility of any professional, be he a lawyer, a doctor, an architect, an accountant or an engineer, to make decisions on behalf of his clients. The responsibility of the professional is to investigate the pros and cons of a given course of action, to weigh the facts and to give a professional opinion to the client allowing the latter to make what is known as "an informed decision". The professional who makes the decision himself will, undoubtedly, be held liable for any additional costs ultimately incurred by the client arising out of a decision which was made in his name and without his input. Perhaps the owner will prefer the cheaper of two roofing systems, in spite of the fact that it may require a more expensive snow-removal program. If that is so, the consultant's role is to describe the alternatives to the owner and to let him make the choice. It is, of course, a good idea to confirm the client's choice in writing.

The theory of informed consent applies for all phases of a professional's mandate from the feasibility study, to the design, to the field services. We note, however, that it is generally during the actual construction process and during the simultaneous rendering of field services that design consultants are inclined to make decisions on behalf of their clients, rather than to allow them to make an informed decision.

CONCLUSION

It is not suggested here that design consultants refuse to cooperate with contractors because they are neither responsible for construction methods nor for site safety. The smooth running of the construction site and the ultimate delivery of an acceptable project to the owner require close cooperation between design consultants and contractors. The purpose of this bulletin is to remind consultants of the delineation of duties and responsibilities, as between contractors and consultants, and to encourage architects and engineers to pass this information along to their field staff. These site representatives or inspectors should arrive on site with a complete understanding of the consultant's mandate for field services and should be made aware that they should not accept responsibilities, and the liabilities ensuing therefrom, which properly belong to the contractor.

Appendix 2

The following *two articles illustrate some aspects of the roll of an expert witness. "The Engineer as an Expert Witness" was prepared by a consulting engineer. The article was originally printed in the December 1981 edition of Civil Engineering and is reprinted with permission of the American Society of Civil Engineers.*

"Cross Examination ... tips for the expert witness" was prepared by a lawyer. The article was originally in the December 1983 edition of the Canadian Consulting Engineer and is reprinted with permission of Southam Publications Ltd.

The engineer as expert witness

Many lawsuits involving responsibility and damages hinge on technical testimony. Here is a summary guide for the prospective expert witness, written from the perspective of a geotechnical specialist, but applicable to all engineers.

JAMES E. HOUGH, P.E., M. ASCE
Principal, James E. Hough and Associates
Cincinnati, Ohio

IT IS ESSENTIAL that the engineer be aware of certain laws, statutes, ordinances, working agreements, and contract agreements and related documents, and how they apply to engineering works.

Increased specialization in technical fields over decades and the increasingly litigious nature of our society have resulted in a great deal of collaboration between lawyers and engineers in litigation involving engineered projects.

Courts of law require that certain facts be established: first, that a duty is owed by one or more of the defendants to the plaintiff (injured party); second, that there was a breach of duty (e.g., negligent construction practices); third, that the breach of duty caused the damage for which relief is sought by the plaintiff; fourth, the extent of the damage. To accomplish this, in cases where the testimony of lay persons is inadequate, expert witnesses are permitted to render opinions based on the facts of the case.

An expert witness can be any person possessing special knowledge, skill, experience, training, or education sufficient to qualify in the subject to which his testimony relates. Such expertise must be established in the courtroom before the witness may testify as expert. The expert witness may then interpret and explain technical facts to enable the court to reach a decision.

Expert testimony is not indisputable and may be controverted by lay testimony establishing inconsistent facts. Since the court considers expert opinion when given as neither more nor less than evidence, a conflict between two experts constitutes a conflict in the evidence.

Your first appearance in court as expert witness likely will be far different from any professional experience you have had. Prior experience in appearing before a meeting of professional peers, at a technical committee hearing or in presenting a technical report will not have prepared you for the probable jolt your ego will undergo while on the stand. If your next trip to the stand is a satisfying experience rather than an ordeal, then the objective of this article will have been fulfilled.

Pre-trial instruments of the court

Deposition. In this relatively informal procedure, usually called by opposing counsel, you are orally examined under oath before the trial to determine the facts in your possession. This is accomplished via questioning by opposing counsel. All questions and answers are recorded by a court reporter. The lawyer with whom you are working also is present. You respond under rules akin to courtroom cross-examination, and what seems to be a straightforward question may have legal implications.

This deposition may be used in the trial to impeach your credibility. Opposing counsel may attempt to trip you up in cross-examination by asking a similar question in a slightly different manner, hoping to obtain an answer different than that given in the deposition. If you are alert to the similarity of the questions, you may ask permission of the court to explain the difference in your answer as being an answer to a different question.

Interrogatory. Comprising a list of questions from opposing counsel requesting answers from you, the line of communication in this procedure is through the lawyer with whom you are working. Normally, you will give answers to the questions to the lawyer who, in turn, will edit and prepare them in proper form and return them to you for corrections. Your answers to the interrogatories will be notarized and may be used in the trial to impeach your credibility.

These procedural tools of discovery enable opposing counsel to "fish" for what you have discovered or concluded in your investigation. Some of the questions may be ambiguous and cannot be fully answered. Should a question be of this sort, your most suitable response would be that the question is ambiguous and cannot be answered. If the question is substantially duplicated by a previous question, your answer should refer by number to that question and response.

How to prepare testimony

1) Withdraw from a case following preliminary appraisal if circumstances and facts appear incongruous and you cannot willingly and ethically support them, or if you have a conflict of interest.

2) Investigate the lawyer before agreeing to work with him. He may be objectionable even though the case is judged worthy.

3) Don't risk your reputation by working with a careless lawyer. An ignorant lawyer is bad, but a careless lawyer is a menace to the profession.

4) Have a definite understanding with the lawyer regarding the need for adequate investigation. Lawyers occasionally want to restrict severely the amount of time you spend in preparation in order to reduce costs.

5) Prepare yourself adequately for the case with the necessary field/laboratory/office investigation. The lawyer mistakenly may believe that technical facts are less than critical to the outcome of the case and that he simply can "out-argue" the opposition.

6) Fully document conditions bearing on the case. Maps, drawings and photographs that clearly demonstrate the facts and your interpretations and conclusions are important. These may be used to explain geologic or geotechnical principles, origin of features, changes induced to preexisting *in situ* conditions by human intervention, changes incurred with time, etc. Three-dimensional drawings or models are particularly beneficial in documenting subterranean conditions and demonstrating their legal relevance to surface features, processes and events.

7) Review your findings with the lawyer well in advance of the court date. A substantial portion of your time may be consumed in educating the lawyer regarding technical matters.

8) Go into court only when certain the lawyer knows all of your findings and conclusions.

9) If there are any skeletons in your closet, e.g., conviction of a crime involving

"Actually, I am familiar with Chapter ten, pages eighty-one to eighty-four of "Foundation Problems and Solutions, Vol. II." In fact, I wrote it."

moral turpitude or inconsistent testimony in prior similar cases, be certain to fully inform the lawyer.

10) Plan your presentation of testimony (in a general way) with the lawyer, preferably 2-4 days prior to the court date. A careful lawyer will prepare an outline and specific questions (with your assistance) so that the relevant facts and your conclusions will be fully affirmed in logical sequence, and your testimony (under oath) will be complete. Remain strictly within the bounds of the case and present data only to elucidate; it is disadvantageous to load testimony with unnecessary information.

When opposing counsel does everything within his power to discredit your testimony, it usually signifies that you have favorably impressed the jury.

11) Take the witness stand only after having carefully prepared yourself. Be prepared to face skilled, and occasionally unfair, cross-examination of your testimony by opposing counsel. This may constitute 5-10 hours of preparation for each hour on the stand. Your preparation should include review of all work and reports bearing on the case, as well as books and any professional reports and published writings of your own that may have pertinence.

12) Be directed by the lawyer and lend him your full support and loyalty.

13) Through discussions with the lawyer, coordinate your testimony with the conclusions of other experts on your team.

14) Study portions of the deposition given by any expert or witness that touches upon the subject of your testimony.

15) Be aware when giving your deposition that opposing counsel may request and obtain copies of notes, correspondence, reports,etc., that you have in your hand at the time of taking the deposition.

16) For background in a lengthy case, study the transcript of relevant court proceedings (where permitted) prior to your appearance.

17) The use of notes to refresh your memory on a point or series of data is permitted while on the stand if reference to them is prefaced by appropriate remarks. The court may ask to inspect your notes in such event.

Following your swearing in, the lawyer (your team leader) will commence his questioning, usually starting with your qualifications and expertise pertaining to the subject of your testimony. After he has sufficiently elicited your qualifications as expert, opposing counsel may cross-examine you regarding your expert qualifications—this is known as *voir dire*. Upon conclusion of the *voir dire* and the court's acceptance of your expert qualifications, the lawyer resumes direct examination, going into the subject of your testimony. When the lawyer has finished his direct examination, opposing counsel may cross-examine you on the subject of the direct examination.

After conclusion of cross-examination, the direct examiner may question you regarding subject matter brought out in the cross-examination. This redirect examination cannot present new subject matter or evidence, but must confine itself to exploration of any facts or contradictions brought forth in the cross-examination.

Upon conclusion of redirect examination, opposing counsel may recross-examine you. This recross-examination is limited to contradictory statements betwe[illegible]n the cross-examinaation and redirect e[illegible]amination. Ordinarily, the examination stops at that point, but the judge may exercise his discretion as to the extent of the examinations. The judge also may engage in reasonable examination of you at any time during the presentation of your testimony.

The tactics of eliciting evidence are those of the lawyer, not the engineer; he usually has definite opinions regarding the sequence of their presentation. This framework often is rather formal and rigid, and not to the liking of the engineer or geologist. Be mindful, however, that court procedure is calculated to elicit the truth in an orderly manner.

When opposing counsel does everything within his power to discredit your testimony, it usually signifies that you have favorably impressed the jury.

Guidelines for testimony—direct examination

1) State clearly and completely (as appropriate) your professional credentials, especially those relevant to the subject of your testimony. Include your education, practical experience, and professional registration or license. Emphasize your professional experience in the geographic area of concern.

2) Your presentation of testimony generally should follow one of two methods. One is to set forth all evidence bearing on the case. The other is to withhold some nonessential evidence which is damaging to the opposition, anticipating that opposing counsel will "rise to the bait" and request it during cross-examination (calculating that you omitted statements on the subject because they contradict your conclusions). This delayed introduction effectively strengthens your testimony and diminishes opposing counsel's enthusiasm for further questioning.

If opposing counsel fails to "take the bait" by not questioning you on the evidence you have chosen to omit, the lawyer may attempt to elicit your statements on the subject during his redirect examination.

3) In opinion testimony, the lawyer will ask if you have an opinion, based upon the facts available to you, regarding some aspect of the matter being litigated, and you will answer. If you respond affirmatively, he then will ask you what that opinion is, and you will give the court your opinion without expanding thereupon. Then he will ask you upon what you base your opinion. Your answer to this question is a most important part of your testimony. Your success will depend largely on two factors: the extent to which you know, understand and look your role; and your ability to present your data clearly and in an easily understood manner.

4) The lawyer occasionally may have difficulty properly phrasing his question. Assist the lawyer by rephrasing his question if the meaning is not precisely stated. You can say: "Do you mean. . . .?" Other times, your use of qualifying words or phrases in answer to a question may compel the lawyer to delve more deeply into the subject. You may force him to ask you to explain if your answer includes "sometimes," "usually," or "under certain circumstances," thereby opening the door for your complete statement.

5) Define and simplify uncommon words and technical words or jargon whenever possible.

6) Present no opinion concerning subjects outside your specific area of expertise, even though the matter may be within your general field of knowledge. A qualified civil engineer expert in geotechnics, for example, may not be qualified to render opinion regarding a structural failure.

7) Display total impartiality. This requires conscious effort. An objective of opposing counsel may be to show prejudice.

8) Keep your eyes on the lawyer and listen carefully thoughout his question; then direct your attention and answer to the jury or judge, not to the lawyer. (He should know the essence of your response before asking.) The case easily can be lost if you lose the attention of the jury (or the judge).

9) Speak in easily audible tones. As you have something worth hearing, speak to the jury (or judge) in an authoritative manner.

10) Try to appear competent yet modest while on the stand. As a jury or judge is inclined to be suspicious of undue assertiveness or arrogance, rely on the presentation of your professional credentials and your demeanor to reveal your authority

on the subject of your testimony.
11) If you do not know the answer to a question, say so.
12) Exaggeration in your response to a question is likely to be a hindrance later in the trial. (An objective of opposing counsel may be to show that your testimony is improbable.)
13) Don't give a cursory "yes" or "no" answer to a complex question. Present the reasons leading to your conclusions if they clarify the basic points in question.
14) Answer only the question asked (if you can), allowing the lawyer to determine the order and presentation of evidence.
15) Don't be reluctant to admit a mistake or to qualify an answer. Your reputation (and the impression you leave with the court) for honesty and sincerity is valuable.

It is foolish to be "clever" in your response to a question. The cross-examiner is performing in his back yard and will have the distinct advantage; if you forget this, he'll show you a few tricks that you may not have heard about. Say as little as possible but as much as necessary.

Guidelines for testimony—cross-examination

1) Opposing counsel will deal with you in one of three ways: (a) as though you do not know your subject or the facts of the case, thereby discrediting you; (b) as though you are unsure about important facts or aspects of the case, thereby discrediting your testimony by eliciting from you conflicting statements for the record; or (c) as though you are well-prepared and truly an expert, in which event ordinarily few questions will be asked for fear of damaging answers.
2) You should never allow your answer to a question to be rushed, although the cross-examiner may try pressuring you to a hasty response. Theoretically, you have unlimited time to answer a question. If your correct answer would require several hours of calculations, so state and await court instructions.
3) Be deliberate and selective, accepting no confusing rapid-fire questions. The cross-examiner gains nothing by asking questions which go unanswered.
4) Do not hesitate, however, if the answer to a question is obvious. A prompt answer is highly effective as it often leaves the jury or judge waiting for the cross-examiner to resume his questioning.
5) The use of compound questions is a common technique of the cross-examiner in order to confuse the expert, jury or judge. Never attempt to answer such questions in a single response; have the cross-examiner choose the one you are to answer.
6) Trick questions are a tool of the cross-examiner. Generally, limit your response to the question asked. You properly may be compelled by the judge to answer "yes" or "no" as the trick question indicates. You may meet the question by responding "Yes (or no) I can explain that," thereby immediately alerting the jury or judge to the attempted trick. Even if opposing counsel avoids your offer to explain, the lawyer with whom you are working will take note when he hears "I can explain that," and he will call forth your explanation on redirect examination. Another approach to trick questions requiring a "yes" or "no" answer is to reply: "I will be happy to answer if the court will allow me to qualify my response."
7) You may be asked if you have talked with anybody about the case. The response may be a statement that you have talked it over at length with the lawyer who called for your presence.
8) The cross-examiner may ask if the attorney told you what to say. Your best answer is that he told you to tell the truth.
9) If asked how much you are being paid to testify, state the amount frankly and matter-of-factly, adding "That is my normal fee."
10) Maintain your composure at all times; just politely smile and be courteous to the cross-examiner and the court. The jury and judge like to see a harassing cross-examiner fail.
11) The cross-examiner may ask if you frequently have differed from other experts. You may answer "Perhaps," adding that you still are convinced that your opinion was correct. (There may be room for such differences in close cases.)
12) It is foolish to be "clever" in your response to a question. The cross-examiner is performing in his back yard and will have the distinct advantage; if you forget this, he'll show you a few tricks that you may not have heard about.
13) Uninformed though the cross-examiner may appear, never underestimate his grasp of the facts. He may be better informed than you are on some point and act uninformed in order to discredit you on a technical matter.
14) If it is true that you have been called upon many times to testify as expert, admit it. The fact that your opinion is much sought after affirms your knowledge and professional competency. And if true, you may add that you are called in consultation very often. Should the cross-examiner ask "How often," thus indicating his inexperience, let him have it (the details).
15) Accept a book or professional paper as authoritative only if you know well its contents relevent to the subject of your testimony. Opposing counsel may have a copy under the table.
16) Do not hestitate to question statements in textbooks by alleged authorities, if you disagree. When you know the book is outdated, inquire as to its copyright date or edition. Your alternative answer may be that textbooks are meant to present principles and so are often not entirely applicable in particular instances. You may also say, simply, "I can explain that." In the latter instance, when asked to explain, be certain that you name the textbooks which are most authoritative and which support your opinion.
17) When the lawyer with whom you are working "objects" to a question, remain silent. Listen carefully to the objection as he may perceive a subtle innuendo in a seemingly innocuous question, thereby alerting you to the dangers of the question. Await the judge's decision whether or not you may answer the question.
18) Say as little as possible but as much as is necessary.

General guidelines

1) Appear in court only when instructed to do so by the lawyer with whom you are working.
2) The jury will be highly critical of your appearance and demeanor; look, dress and conduct yourself accordingly.
3) Closely follow court procedure and the rules of evidence. Although they appear inexplicably restrictive, each has an important relation to the just determination of the controversy.
4) It usually is unwise to discuss any aspect whatsoever of the case in the corridor or (during recess) in the courtroom, except with the lawyer with whom you are working. Opposing counsel may have posted clerks near you, using what is overheard to your detriment on cross-examination.
5) Your effusive greeting of an opposing expert witness likely will attract attention, thereby enhancing his image. Even though he may be a friend or a colleague, you are obligated to make little of him and to destroy his opinion (rightfully) by your superior opinion.
6) Never consult with an opposing expert.
7) Promptly leave the courtroom upon completion of your testimony, unless otherwise requested by the lawyer. ▽

James E. Hough has had 27 years of experience as an engineering geologist, soil and foundation engineer, and geotechnical engineer in both the private and public sectors. He has published extensively on the subject of landslides. His most recent publication is Engineering Geology of the Cincinnati Area, published by the Geological Society of America.

Cross-examination ... tips for the expert witness

By J. Frederick Sagel

We live in an era that looks favorably upon experts, partly due to television and the impact of the space age. People are deluged with expert advice. In the courtroom, jurors readily accept the idea that a qualified expert must know all there is to know about the subject at issue.

Since engineers are often expert witnesses, it is important to know how lawyers relate to experts, how important the engineering expert may be, how important it is for lawyers to brief experts properly.

Communication vital

Shakespeare said it better than most of us. You will recall that in Henry VI Part II, Dick Butcher said, "The first thing we do, let's kill all the lawyers." And old Cade replied, "Nay, that I mean to do." That expresses the attitude of most experts.

I have never met an expert of any kind who has any liking for lawyers. You see, the expert knows he is an expert. Take a medical doctor - everyone defers to him. Everyone thinks he is a great fellow except that lawyer who will ask him questions about things he studied in medical school and forgot years ago, things he knows are not germane to the problem. The doctor worries about this. He worries, too, because one of his colleagues told him that the lawyer will ask him how much money he makes and whether he declared all of it in his tax return.

Briefing the witness

Whether you are being called upon in defence of your own case or an expert witness, it is important that the lawyers detail the facts of the case for you. An engineering expert, for example, may have learned all he needs to prepare his report for examining the plaintiff, but he also needs to know facts of the case. It is important that you are briefed thoroughly on all of the evidence. An expert dreads being caught off guard in the witness box. There is no need for it. If you are called upon as an expert and you walk into a trap and you are subjected to ridicule, there is only one person to blame - the lawyer who hired you and who was supposed to protect you.

As an engineering expert, it is best that you are brought into the case gradually. You should feel relaxed about testifying in court. You should know that if you can give a strong opinion in your report, there is every chance that the case will be settled. As well, you should be encouraged to be on the lawyer's side, which, of course, is the right side.

Many people are convinced that the expert who really persuades a jury is the independent, objective, non-articulate type. Personally, I strongly disagree with this. I would go into a lawsuit with an objective, uncommitted, independent expert about as willingly as I would occuy a foxhole with a couple of non-combatant soldiers.

If the expert chosen for the case is independent and not firmly committed to the theory you want, it is extremely harmful to put him on the stand. You cannot be sure of his answers on cross-examination. An expert on the witness stand must know which side he and the lawyer who retained him are on.

The trial lawyer must make of the expert a convincing, persuasive witness. The lawyer deals in words and he knows how to put the package together to impress the jury or the judge favorably. It is his job to instruct the expert, although this is an exercise that often requires great tact and firm conviction.

One might say to an expert, "Doctor, you are an expert in medicine, busy every day seeing patients with interesting ailments, but we have been in court every day for years and know what approach will interest jurors. Understand that we are not attempting to influence you in this case. We both know that we wouldn't do that and you wouldn't allow it. We are saying only that the way you express yourself — the image you project to the jury — is vital to our case. Above all, don't get angry and bristle at the cross-examiner. If he asks you a sneaky question, just look at the jurors. If you recognize the questions as crooked, they will too."

Coping with cross-examination

I like to suggest specific answers to experts for the cross-examiner who tries to discredit him by insinuating that the expert has become a jprofessional witness. For instance, when they say to you as an engineer, "You come to court a lot don't you?" You might answer, "Well, I don't know what you mean by a lot; unfortunately I'm here a lot more than I would like to be."

"Isn't it a fact that you testify for Mr. Lilly and Mr. Sagel at least once a month?"

"No, I don't think that's a fact. I've been in court before when Mr. Lilly and Mr. Sagel were involved in a case, but not once a month. No, no, that's exaggerating, counsel."

"How many times have you been in court for Mr. Lilly and Mr. Sagel?"

"Well, the last time I was in court, it seems to me that Mr. Lilly was trying to play the part you are. He represented the fellow who was trying to make something out of nothing."

At this point, I immediately say, "I move that the answer be stricken." In any case, you can see how that gets the witness into danger.

The expert can only be effective if he is given all of the information he needs. It is most important that you know how valuable you are to a case. You must feel like part of the team and you should be involved in the total lawsuit.

I'm sure you have been asked on cross-examination: "Well, isn't this possible?" As a lawyer, you would like the expert to say no, but instead he answers. "Anything is possible, you might have something." Engineers must learn that phrase at school. This situation is less likely to arise if the engineer knows all the proof. His testimony can be as positive as the proof I'm going to be offering.

This tactic was used in a case where we were defending against a plaintiff who claimed he was going blind. Before the trial, we showed our medical expert a film of the plaintiff doing things that a man with his claimed disability would never be able to do. We then asked the experts if they could answer confidently when the opposing lawyer asked if the plaintiff could have the claimed injury. The experts were able to answer with stunning certainty that it was not possible, knowing that we would later show the film which would prove them right. But you can only do that if you are well acquainted with case, and it's up to your lawyer to do that for you.

I have found that engineering specialists are better able to testify in a complicated case if I get some preliminary work done with them, if I define the terms with them, go through the case and show them what other evidence we have. In that way, both the lawyer and the engineer are totally familiar with what's happening.

A lawyer's duty

You as an expert have to know that you have a fighting lawyer on your side. You don't want a lawyer who thinks the witness can take care of himself. This is entirely wrong. You as an expert should be confident that when a cross-examiner asks you what your yearly income is, the lawyer will object to the question even before you begin to answer it. It is the lawyer's job to protect the expert from all irrelevant questions. When a question, on the other hand, is germane to the lawsuit and involves the witness's expertise, the lawyer can't help you. However, he should have briefed you well enough that you can answer with confidence.

Plain language

Engineering evidence is difficult for a judge or a jury to understand. If you are being sued or, if you are acting as an expert, it is important to make things as simple as possible. The language used by the interrogator and witness must be as plain as possible. The tryers of fact must understand the questions and answers. To accomplish this objective, a translation of terms into simple language is usually necessary. It's important for everyone in a court to know in non-technical language what the witness means.

The hidden punch

A layman who thinks he knows more about cross-examination than any lawyer, is the Perry Mason fan who sees the witness finally break down under Perry's incisive cross-examination and admit that it was he who perpetrated the crime and that Perry's client should go free.

Unfortunately, as all lawyers know, cross-examination does not work that way. Indeed, the final question may be the one that hurts the most.

This anecdote I heard from a defence trial lawyer recently illustrates the point. He warned of the dangers of

"You as an expert witness have to know that you have a fighting lawyer on your side. You don't want a lawyer who thinks the witness can take care of himself."

cross-examination with the story of the witness on the stand, an old Amish farmer, who had been driving down a road in his surrey. A speeding car missed rounding a turn, rammed into the surrey, turned it over and knocked the horse and the farmer into a ditch. On cross-examination, the lawyer said, "Now sir, when this happened, the defendant came over to talk to you. Isn't that true?"

"Yes", the farmer replied.

"And at the time, you said you weren't hurt. Isn't that right?"

"Yes," the farmer agreed, despite the fact that on direct examination he had taken considerable trouble to detail each and every one of the terrible injuries he had suffered.

Instead of letting well enough alone, the lawyer asked a final question. "Why did you say you were not hurt?"

"Well", the farmer answered, "When his car hit my horse, it knocked him over into the ditch. My horse was on his back with his feet kicking. He had a broken leg. I was in the ditch and the buggy was turned over. The defendant got out, looked at the horse, went back to his car, took a pistol from the glove compartment, came back, put it to the horse's head and shot him. Then he came and stood over me with the smoking pistol and said, "Are you hurt?" I said, "No sir!"

Tactics &tips

All of us who have tried lawsuits have asked what we thought were innocent questions, only to discover that the answer had a hidden punch capable of giving us a bloody nose. Since none of us enjoys the sight of our own blood, it is important to take the proper steps in preparation for cross-examination. Preparing an integrated lawsuit planned thoroughly from beginning to end is the basis of avoiding the surprise punch scenario. The tactics of cross-examination should fit into the total plan. As the above anecdote illustrates, it is important that a lawyer should not ask a question to which he does not know the answer.

Proper cross-examination depends upon totally adequate preparation and a thorough understanding of your purpose. A lawyer doesn't get up to ask questions off the top of his head. There is a goal in mind, for instance; to test the voracity of your witness, to establish factual concessions that will establish a defence as a matter of law, to obtain material for a closing argument, or to fortify and corroborate testimony of witnesses or show a conflict in the facts given by your opponent's witness.

Cross-examination, I have found, is an art rather than a science. It can be learned only through long personal practice. It is especially difficult when you are taking on someone who is an expert, such as an engineer. You are into his field, and his field alone. You are an expert in words, but he is an expert in what he is talking about, so it is most important to be careful.

There are some things I think lawyers should not do with the expert witness. It might be useful to keep these points in mind the next time a lawyer is going over your testimony with you.

- Lawyers should not over-examine. My story about the Amish farmer serves to illustrate that point.
- They should not ask wide open questions. It is important to keep a tight rein on the witness. When wide open questions are asked, the witness has the run of the lawsuit. It is important, on cross-examination, to let the lawyer run it. If the witness is asked a wide open question, he starts running it and the lawyer's purpose will be destroyed.
- The lawyer shouldn't use sledge hammer techniques when he examines children or old ladies. Every courtesy should be extended to all witnesses.
- The lawyer shouldn't repeat questions which were asked on direct examination. To have the witness repeat his testimony on direct examination will only help the opponent's case.
- The lawyer shouldn't ask any question to which he doesn't know the answer. He doesn't know how badly it will hurt him. He should avoid the why question, because it is totally unpredictable. To accomplish this, he should frame his questions with you, in advance so that there is only one answer possible.
- The lawyer shouldn't examine at all if no purpose can be achieved. It's often very effective to say, "No questions"

- The lawyer shouldn't attempt to outdo the expert witness in his field by using the vocabulary of the art in his examination. Every engineer should be approached with lay language rather than with technical phraseology.

Briefing witnesses is a lawyer's vital duty.

Witness take note

There are also things the expert witness should watch for, to avoid being hurt. For instance:

- One should be careful of one's past background. If you are going into the stand, either as an expert or as a person who is going to be sued, you are putting your professional reputation on the line. Tell your lawyer if you have any flaws in that reputation. If you don't, the chances are that the other lawyer might have found them and will cross-examine you about them. It is most embarrassing to show that you did poorly in school in an area in which you are giving expert testimony. It is also embarrassing to show that you failed certain courses or had other problems in your career. Lay it on the line to your lawyer so that he can adequately protect you in advance.
- If you are appearing in an examination, every article that you have ever written can be cross-examined upon. If you have written an article to the contrary of what you are saying in the stand, be careful. You may be cross-examined on it. Lay that on the line to your lawyer as well.
- If you are being sued, you will have gone through examinations for discovery. These are pre-trial examinations where the other lawyer asks you questions if you are a party to the lawsuit. These examinations can often be long and tedious. They will be used against you, however, at trial. Read them carefully and make sure there isn't any confict between what you have said on the examinations and what you say at trial.
- Watch the trick question about being paid. If you are being called as an expert, you shouldn't hedge about the fact that you are being paid. Be forthright and honest about it. The judge knows that you have; to make a living if you are being called as an expert, and there is nothing embarrassing to admitting it.
- Be careful about conflicting evidence. If you put forth proposition, make sure that it doesn't contradict an opinion that is widely held and comes out of an engineering textbook. The other lawyer will ask you if your theory is true and will then put the theory in the textbook to you and ask you why your theory differs with that of the textbook. If you are going to do such a thing, tell your counsel in advance and have a good explanation.

J. Frederick Sagel is Counsel for a Toronto legal firm with extensive experience in the area of litigation related to engineering.

Appendix 3

The code

of ethics of the Association of Professional Engineers of British Columbia is typical of those for other Provincial and State engineering associations.

Code of Ethics

14 Preamble – The following is prescribed as the Code of Ethics of the Association, and the Engineer is bound by its provisions just as he is bound by the provisions of the Engineers Act, 1979 and by the Bylaws of the Association.

The professional engineer shall act at all times with fairness, loyalty and courtesy to his associates, employers, employees and clients, and with fidelity to the public needs. He shall approach his work with devotion to high ideals, personal honour and integrity.

The purpose of the Code is two-fold:

(1) To give general statements of the principles of honourable conduct which, over the years, members of the profession of engineering have come to accept as required of each member in order that he may fulfill his duty to the public, to the profession, and to his fellow members.

(2) To give some specifics in the sub-sections, both of required standards and prohibited actions, in order that they may act as a guide, to the intent of the general statements. These specifics, it is emphasized, are only some examples of the broad principles upon which members of this profession must appraise and govern their own conduct.

The following Code of Ethics is promulgated as a general guide and not as a denial of the existence of other duties equally imperative, but not specifically included.

Section 1

The Engineer will be guided in all his professional relations by the highest standards of integrity.

(a) He will be realistic and honest in the preparation of all estimates, reports, statements and testimony.

(b) He will not distort or alter facts in an attempt to justify his decisions or avoid his responsibilities.

(c) He will advise his client or employer when he believes a project will not be successful or in the best interests of his client or his employer or the public.

(d) He will not engage in any work outside his salaried work to an extent prejudicial to his salaried position.

(e) In the interpretation of contract documents, he will maintain an attitude of scrupulous impartiality as between parties and will, as far as he can, ensure that each party to the contract will discharge the duties and enjoy the rights set down in the contract agreement.

(f) He will not use his professional position to secure special concessions or benefits which are detrimental to the public, his clients or his employer.

Section 2

The engineer will have proper regard for the safety, health and welfare of the public in the performance of his professional duties. He will regard his duty to the public safety and health as paramount.

(a) He will guard against conditions that are dangerous or threatening to life, limb or property on work for which he is responsible, or if he is not responsible will properly call such conditions to the attention of those who are responsible.

(b) He will present clearly the consequences to be expected if his engineering judgment is overruled.

(c) He will seek opportunities to work for the advancement of the safety, health and welfare of his community.

(d) He will guard against conditions which are dangerous or threatening to the environment and he will seek to ensure that all standards required by law for environmental control are met.

Section 3

The Engineer may promote and advertise his work or abilities provided that:

(a) The advertising preserves the public interest by reporting accurate and factual information which neither exaggerates nor misleads.

(b) The advertising does not impair the dignity of the profession.

(c) Statements do not convey criticism of other engineers directly or indirectly.

Section 4

The Engineer will endeavor to extend public knowledge and appreciation of engineering and its achievements and will endeavor to protect the engineering profession from misrepresentation and misunderstanding.

(a) He will not issue statements, criticisms, or arguments on engineering matters connected with public policy which are inspired or paid for by private interests, unless he indicates on whose behalf he is making the statement.

Section 5

The Engineer may express an opinion on an engineering subject only when founded on adequate knowledge and honest conviction.

(a) In reference to an engineering project in a group discussion or public forum, he will strive for the use of pertinent facts, but if it becomes apparent to the engineer that such facts are being distorted or ignored, he should publicly disassociate himself from the group or forum.

Section 6

The Engineer will undertake engineering assignments for which he will be responsible only when qualified by training or experience; and he will engage, or advise engaging, experts and specialists whenever the client's or employer's interests are best served by such service.

(a) He will not sign or seal plans, specifications, reports or parts thereof unless actually prepared by him or prepared under his supervision.

Section 7

The Engineer will not disclose confidential information concerning the business affairs or technical processes of any present or former client or employer without his consent.

Section 8

The Engineer will endeavor to avoid a conflict of interest with his employer or client, but when such conflict is unavoidable, the Engineer will fully disclose the circumstances to his employer or client.

(a) He will inform his client or employer of any business connections, interests, or circumstances which may be deemed as influencing his judgment or the quality of his services to his client or employer.

(b) He, while a member of any public body, will not act as a vendor of goods or services to that body.

Section 9

The Engineer will uphold the principle of appropriate and adequate compensation for those engaged in engineering work.

(a) He will not normally undertake or agree to perform any engineering service on a free basis, except for civic, charitable, religious, or nonprofit organizations when the professional services are advisory in nature.

(b) He will not compete improperly by reducing his usual charges to underbid a fellow engineer after having been informed of that engineer's charge.

Section 10

The Engineer will not accept compensation, financial or otherwise, from more than one interested party for the same service, or for services pertaining to the same work, unless there is full disclosure to and consent of all interested parties.

(a) He will not accept financial or other considerations, including free engineering designs, from material or equipment suppliers as a reward for specifying their product.

(b) He will not accept commissions or allowances, directly or indirectly, from contractors or other parties dealing with his clients or employer in connection with work for which he is responsible.

Section 11

The Engineer will not compete unfairly with another engineer by attempting to obtain employment or advancement or professional engagements by taking advantage of a salaried position, or by criticizing other engineers or by other improper or questionable methods.

(a) He will not attempt to supplant another engineer in a particular employment after becoming aware that definite steps have been taken toward the other's employment.

(b) He will not offer to pay or agree to pay either directly or indirectly, any commission, political contribution, gift, or other consideration in order to secure work.

(c) He will not solicit or accept an engineering engagement on a contingent fee basis if payment depends on a finding of economic feasibility or other preconceived conclusion.

Section 12

The Engineer will not attempt to injure maliciously or falsely, directly or indirectly, the professional reputation, prospects or practice of another person.

(a) He will not accept any engagement to review the work of a fellow engineer except with the knowledge of and after communication with such fellow engineer, where such communication is possible.

(b) He will refrain from expressing publicly an opinion on an engineering subject unless he is informed as to the facts relating thereto.

(c) Unless he is convinced that his responsibility to the community requires him to do so, he will not express professional opinions which reflect on the ability or integrity of another person or organization.

(d) He will exercise due restraint in his comments on another engineer's work.

(e) If he considers that an engineer is guilty of unethical, illegal or unfair practice, he will present the information to the Registrar of the Association.

(f) An engineer is entitled to make engineering comparisons of the products offered by various suppliers.

Section 13

The Engineer will not associate with or allow the use of his name by an enterprise of questionable character, or by one which is known to engage in unethical practice.

(a) He will not use association with a non-engineer, a corporation, or partnership as a "cloak" for unethical acts, but must accept personal responsibility for his professional acts.

Section 14

The Engineer will give credit for engineering work to those to whom credit is due, and will recognize the proprietary interests of others.

(a) Whenever possible, he will name the person or persons who may be individually responsible for designs, inventions, writings, or other accomplishments.

(b) When an engineer uses designs supplied to him by a client or by a consultant, the designs remain the property of the client or consultant and should not be duplicated by the engineer for others without express permission.

(c) Before undertaking work for others in connection with which he may make improvements, plans, designs, inventions, or other records which may justify copyrights or patents, the engineer should enter into a positive agreement regarding the ownership of such copyrights and patents.

Section 15

The Engineer will co-operate in extending the effectiveness of the profession by interchanging information and experience with other engineers and students, and will endeavor to provide opportunity for the professional development and advancement of engineers in his employ or under his supervision.

(a) He will encourage his engineering employees in their efforts to improve their education.

(b) He will encourage engineering employees to attend and present papers at professional and technical society meetings.

(c) He will urge his qualified engineering employees to become registered.

(d) He will assign a professional engineer duties of a nature to utilize his full training and experience, insofar as possible.

(e) He will endeavor to provide a prospective engineering employee with complete information on working conditions and his proposed status of employment, and after employment will keep him informed of any changes in them.

Section 16

The Engineer will observe the rules of professional conduct which apply in the country in which he may practise. If there be no such rules, then he will observe those set out by this code.

Interpretation

15 In the event of any dispute as to the meaning or intent of these Bylaws, the interpretation of the Council shall be final, subject to the right of Appeal as contained in Section 32 of the Act.

Where the word "Act" appears in the foregoing Bylaws, it shall include the Engineers Act and all subsequent Amending Acts, unless the context otherwise requires.

Repeal of Old Bylaws

16 Upon the coming into force of the foregoing Bylaws, all the Bylaws of the Association previously in force shall stand revoked.

Appendix 4

The following *Professional Liability Insurance Policy is reprinted with permission of the Simcoe & Erie General Insurance Company. This policy is endorsed by the Association of Consulting Engineers of Canada and the Royal Architectural Institute of Canada.*

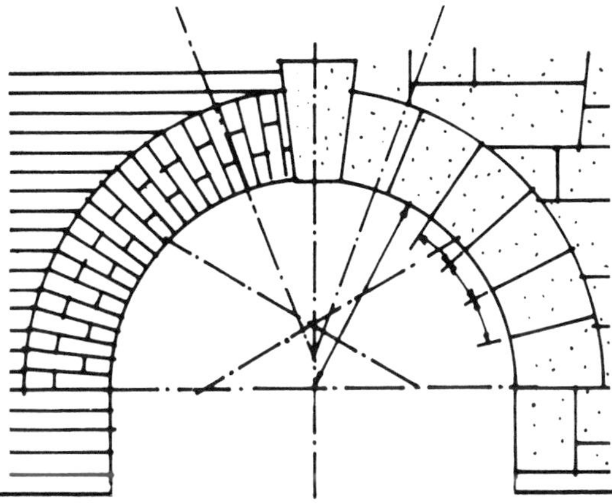

PROFESSIONAL LIABILITY INSURANCE PROGRAM

Simcoe and Erie General Insurance Company

Master Policy L55595

NPA

THE INSURER

THE INSURANCE MANAGERS

THE AGREEMENTS BETWEEN YOU AND THE INSURER

This POLICY is the contract between YOU and THE INSURER and was issued in consideration of the premium YOU have agreed to pay. This POLICY was issued in reliance upon the representations which YOU have made and on the insurance application which is now part of this contract.

PART I DEFINITIONS

Throughout this POLICY certain words have been capitalized to indicate that they have a specific meaning as shown below:

YOU, YOUR
The NAMED INSURED shown on the Declarations Page of this POLICY and all other persons or organizations defined as 'INSURED' below.

THE INSURER
YOUR insurance company, Simcoe and Erie General Insurance Company.

NAMED INSURED
The person(s) or organization(s) specifically designated on the Declarations Page of this insurance certificate.

POLICY
The insurance certificate issued to the NAMED INSURED which certificate shall be considered a separate and individual POLICY incorporating the terms and conditions of the Master Policy L55595, the application for insurance, the Declarations Page and any endorsement issued by THE INSURER.

INSURED
1. The NAMED INSURED, and
2. any present or former partner, executive officer, director, shareholder or employee of the NAMED INSURED while acting within the scope of his duties for the NAMED INSURED.

POLICY PERIOD
The period from the Inception Date of this POLICY to the Expiration Date both as shown on the Declarations Page or such lesser period as a result of the cancellation of this POLICY.

CLAIM
1. A written or a verbal demand for money or services, or
2. a written or verbal allegation of breach in the rendering of or failure to render professional services

received by YOU and resulting from a single error, omission or negligent act.

All CLAIMS arising from a single error, omission or negligent act shall be considered a single CLAIM regardless of the number of INSUREDS or the number of persons or organizations making a CLAIM.

CLAIM EXPENSE
All the expenses THE INSURER incurs to investigate, defend, settle, arbitrate or litigate a CLAIM covered by this POLICY. This includes costs and fees for the hiring of investigators, adjusters, experts, consultants, arbitrators, mediators and lawyers and also court and arbitration costs and costs for the attendance of witnesses other than YOU.

DAMAGES
Compensatory DAMAGES payable to claimants but does not include fines, penalties (whether contractual or other), punitive or exemplary DAMAGES, or fees which have either not been paid to YOU or which YOU are asked to return.

DEDUCTIBLE
The first portion of the DAMAGES payable by YOU for each CLAIM and which amount YOU have agreed to pay in consideration for a reduced premium for this POLICY.

THE INSURER agrees that YOU shall not be required to pay more than twice the amount shown as the DEDUCTIBLE on the Declarations Page of this POLICY for CLAIMS concerning which THE INSURER has paid damages under this POLICY. If the POLICY PERIOD is longer than one year, then YOU will be responsible for twice the amount shown as DEDUCTIBLE for each twelve (12) month period, or part thereof, counting from the inception date shown on the Declarations Page.

PART II
YOUR INSURANCE COVERAGE

THE INSURER'S OBLIGATIONS

THE INSURER is formally undertaking to fulfill three (3) obligations for YOUR benefit. YOUR POLICY DEDUCTIBLE applies to the first obligation only. The second and third obligations are covered with no DEDUCTIBLE.

1. DAMAGES

In excess of YOUR DEDUCTIBLE, THE INSURER will pay on YOUR behalf all sums which YOU become liable to pay as DAMAGES arising out of a CLAIM providing YOUR liability is the result of an error, omission or negligent act in the performance of professional services for others in YOUR capacity as an architect or engineer.

The maximum amount THE INSURER will pay as DAMAGES for each CLAIM, no matter how many INSUREDS there are under this POLICY or how many persons or organizations make a CLAIM, and the aggregate amount of liability for all CLAIMS made against YOU during the POLICY PERIOD, are as shown on the Declarations Page of this POLICY.

2. DEFENCE

THE INSURER will defend YOU in any civil suit or arbitration proceedings, for which coverage is provided by this POLICY, even if the allegations against YOU are groundless, false or fraudulent. THE INSURER will conduct such investigation and negotiations as it deems expedient. THE INSURER'S obligation to defend YOU cease as soon as YOU have exhausted its limits of liability through payment of DAMAGES.

3. SUPPLEMENTARY PAYMENTS

Until YOU have exhausted THE INSURER'S limits of liability, THE INSURER will pay, for each CLAIM in excess of its limits of liability the following:

(a) CLAIM EXPENSES;

(b) all premiums on appeal bonds or bonds to release attachments. THE INSURER has no obligation to furnish such bonds but only to pay the premiums thereon;

(c) all costs taxed against YOU, all court and arbitration costs owed by YOU and all interest upon that part of a judgement which falls within the remaining limits of liability at the time;

(d) YOUR expenses incurred for emergency medical and surgical relief to others and which YOU deemed necessary following an accident which YOU honestly believed might have been the result of an error, omission or negligent act on YOUR part.

YOUR POLICY TERRITORY

THE INSURER'S obligations under this POLICY apply to CLAIMS arising out of actual or alleged errors, omissions or negligent acts which occur anywhere in the world provided CLAIMS are made and proceedings are instituted in Canada, or in the United States of America, its territories and possessions.

YOUR POLICY PERIOD

YOUR POLICY covers each CLAIM made against YOU for the first time during the POLICY PERIOD no matter when the actual or alleged error, omission or negligent act took place. There are three conditions which must be met for such a CLAIM to be covered. First, YOU must have reported the CLAIM to THE INSURER during the POLICY PERIOD. Secondly, YOU must have had no knowledge, prior to the POLICY PERIOD, of such CLAIM or of the circumstances, dispute or controversy out of which it arises. Thirdly, there must not be any other valid and collectible insurance available to YOU concerning such CLAIM.

Also, for your protection, if during the POLICY PERIOD YOU report to THE INSURER circumstances of an error, omission or negligent act which any reasonable person or organization would expect to subsequently give rise to a CLAIM, then THE INSURER will consider these a CLAIM even if a formal demand is advanced against YOU only after the POLICY PERIOD.

Any such CLAIM shall be subject to the limit of liability and DEDUCTIBLE in effect at the time the circumstances were reported to THE INSURER.

THE INSURER'S LIMITS OF LIABILITY

The maximum amounts THE INSURER will pay as DAMAGES per CLAIM and for the entire POLICY PERIOD are as shown on the Declarations Page of this POLICY no matter how many INSUREDS there are under this POLICY or how many persons or organizations make a CLAIM.

If the POLICY PERIOD is longer than one year, THE INSURER'S limits of liability as shown on the Declarations Page of this POLICY shall apply separately to each twelve (12) month period or part thereof counting from the Inception Date also shown on the Declarations Page.

THE INSURER'S obligations to defend and to make supplementary payments are in addition to its limits of liability as spelled out in YOUR POLICY.

PART III
THE EXCLUSIONS
TO YOUR INSURANCE COVERAGE

EXCLUSIONS

1. THE INSURER will not cover YOU, pay DAMAGES, provide YOU with a defence or make supplementary payments for CLAIMS arising out of:

   (a) the infringement of any trademark or patent or copyright;

   (b) YOUR insolvency or bankruptcy or YOUR undergoing receivership or liquidation;

   (c) YOUR advising or requiring, or failure to advise or require, any form of insurance, suretyship or bond;

   (d) YOUR failure to complete drawings, plans, specifications or schedules on time or YOUR failure to act upon shop drawings on time, unless such failure is the result of an error or inaccuracy in the preparation of these documents;

   (e) the liability of others YOU have assumed under contract or agreement except that THE INSURER will cover YOU for YOUR liability for YOUR employees, agents, servants and subconsultants;

   (f) express warranties, guarantees and penalty clauses YOU have given for the benefit of others unless YOUR liability would have already existed at law in the absence thereof;

   (g) the performance of services not usual or customary for professional architects or engineers;

   (h) estimates of profit, return on capital, economic return or other estimates giving rise to forecasts of economic return;

   (i) YOUR participation in a joint venture, partnership, associateship or any other entity which has not been endorsed on this POLICY as an additional NAMED INSURED unless this POLICY was specifically issued for this purpose;

   (j) the nuclear energy hazards as defined in the nuclear energy exclusion endorsement forming part of this POLICY.

2. THE INSURER will not cover YOU, pay DAMAGES or provide YOU with a defence or make supplementary payments for CLAIMS made against YOU:

   (a) by a business enterprise

      (i) in which YOU either directly or indirectly have an interest; or

      (ii) that directly or indirectly has an interest in YOU;

   (b) by any employee, director, partner or officer of any such business enterprise.

PART IV
GENERAL CONDITIONS

YOUR DUTIES IN THE EVENT OF A CLAIM

WHAT YOU MUST DO

1. **NOTICE:** As soon as YOU become aware of a CLAIM, YOU must immediately notify THE INSURER, giving all pertinent details as to the circumstances surrounding the CLAIM. As events unfold which may have an effect on the CLAIM, YOU must continue to keep THE INSURER informed.
2. **COOPERATION:** YOU must cooperate with THE INSURER and, upon request, provide written statements, submit to examinations and questioning, assist in effecting settlement, secure and give evidence and assist in any reasonable way THE INSURER deems necessary. YOU must give this cooperation at YOUR own cost.
3. **DEDUCTIBLE:** YOU must pay YOUR DEDUCTIBLE promptly upon request.

WHAT YOU MUST NOT DO

1. **ADMISSIONS:** YOU must not admit responsibility, assume any obligation or make any commitment of money or services without THE INSURER'S consent, even if you believe there may have been an error, omission or negligent act on YOUR part. Any such admission, obligation or commitment will vitiate this POLICY as far as that particular CLAIM is concerned. The only exception to this is the cost of emergency medical or surgical relief to others YOU have incurred in good faith.
2. **RECOVERIES:** YOU must not do anything which will imperil THE INSURER'S rights of recovery against any other party.

YOUR CONSENT TO SETTLE

THE INSURER will not settle any CLAIMS without the consent of the NAMED INSURED.

If the NAMED INSURED refuses to consent to the settlement of a CLAIM as recommended by THE INSURER, then all THE INSURER'S obligations with respect to that CLAIM shall cease. If later YOU settle the CLAIM, or if the matter is resolved through arbitration or litigation, then THE INSURER'S liability for that CLAIM shall not exceed the amount for which the CLAIM could have been settled plus the costs and expenses incurred up to the date of refusal of consent.

OTHER INSURANCE

If on the Declarations Page of this POLICY it is indicated that this is a General Practice Policy, then this insurance shall be in excess of all other valid and collectible insurance available to YOU and this insurance shall not be called upon in contribution.

If on the Declarations Page of this POLICY it is indicated that this is a Specific Project or Joint Venture Policy, then this insurance shall be primary to any other professional liability insurance.

THE INSURER'S RIGHTS TO RECOVER FROM OTHERS

After THE INSURER has paid DAMAGES under this POLICY, YOUR rights to recover against any other party are automatically transferred to THE INSURER to the extent of the payment it made. YOU shall do everything needed to assist THE INSURER and YOU must not prejudice its rights of recovery.

ASSIGNMENT OF POLICY

YOU cannot assign YOUR rights under this POLICY to anyone else without THE INSURER'S consent. If YOU should be adjudged bankrupt, insolvent, incompetent or die during the POLICY PERIOD, this POLICY will cover YOUR legal representatives in the same manner as it presently covers YOU.

YOU agree that any notice of any kind THE INSURER mails to the NAMED INSURED at the address shown on the Declarations Page shall constitute notice to YOUR legal representatives.

MORE THAN ONE INSURED

If there is more than one NAMED INSURED under this POLICY, this POLICY must be read as if a separate POLICY had been issued to each. This will not, however, increase THE INSURER'S limits of liability.

THE INSURER'S RIGHT OF AUDIT

During the POLICY PERIOD, during any extension thereof and for one year thereafter, THE INSURER has the right to inspect YOUR premises and operations and to examine and audit YOUR books, but strictly as they relate to this insurance or to the calculation of the premium for this POLICY. THE INSURER assumes no responsibility whatsoever by exercising or declining to exercise such right.

PREMIUM

YOUR premium for this POLICY will be shown on the Declarations Page as either a Fixed Premium or a Deposit Premium adjustable upon cancellation or expiry of this POLICY.

If YOUR actual POLICY Premium is an Adjustable Deposit Premium, then upon cancellation or expiry of this POLICY, YOU must declare to THE INSURER the amount of gross fees for professional services billed by YOU during the POLICY PERIOD. THE INSURER will then calculate YOUR final premium by multiplying the rate per one hundred dollars ($100.00) of fees shown on the Declarations Page by the total amount of fees YOU have declared.

If this premium adjustment produces a difference between the final premium and the deposit premium of less than two hundred dollars ($200.00), YOU and THE INSURER both agree to waive the adjustment and forgive either the additional premium payable by YOU or the return premium payable to YOU as the case may be. Any premium adjustment is subject to THE INSURER retaining at least the Minimum Retained Premium shown on the Declarations Page.

Gross fees shall mean the gross amount billed by YOU to clients, including that portion of fees which YOU pass on to subconsultants but excluding charges for extraordinary disbursements. Gross fees shall also include the market value of non-monetary compensation received by YOU in lieu of fees for professional services rendered.

CANCELLATION BY YOU

YOU may cancel this POLICY at any time by giving THE INSURER notice in writing stating the date cancellation is to take effect. YOU must return the POLICY to THE INSURER if possible.

CANCELLATION BY THE INSURER

If THE INSURER decides to cancel this POLICY, it must notify the NAMED INSURED, in writing, stating the date cancellation is to take effect.

If THE INSURER cancels because YOU have not paid the entire premium, the date of cancellation shall be at least fifteen (15) days after the date on which the notice was mailed.

If THE INSURER cancels for any other reason, the date of cancellation shall be at least forty-five (45) days after the date on which the notice was mailed.

If following cancellation there is a return premium payable to YOU, THE INSURER'S cheque will be sent to YOU as soon as possible but the cancellation is not contingent upon this.

PREMIUM ADJUSTMENT FOR CANCELLATION

If the premium for this POLICY is a Fixed Premium, the amount of earned premium calculated upon cancellation shall be computed on a pro-rata basis in accordance with the number of days during which the POLICY was in force.

If the premium is an Adjustable Deposit Premium, the amount of earned premium shall be calculated by multiplying the rate per one hundred dollars ($100.00) of fees shown on the Declarations Page by the amount of gross fees for professional services billed by YOU while the POLICY was in force.

THE INSURER is entitled to retain the Minimum Retained Premium shown on the Declarations Page.

NOTICE TO EACH OTHER

The NAMED INSURED shall be considered the agent of all other INSUREDS under this POLICY.

All notices THE INSURER sends to YOU under this POLICY must be sent to the NAMED INSURED at the address shown on the Declarations Page.

All notices YOU send to THE INSURER under this POLICY must be sent to either:

Simcoe & Erie General Insurance Company
505 York Blvd.,
Hamilton, Ontario
L8N 3S3

or

National Program Administrator
222 Queen Street, 4th Floor
Ottawa, Ontario
K1P 5V9

POLICY CONFORMITY WITH STATUTES

Terms of this POLICY which are in conflict with the statutes of the province wherein this POLICY is issued are hereby amended to conform to such statutes.

APPENDIX IV

SPECIAL ENDORSEMENT NUCLEAR ENERGY EXCLUSION

Endorsement No. 1

This POLICY does not apply to injury, sickness, disease, death, damage or destruction

(a) with respect to which an INSURED under this POLICY is also insured under a contract of nuclear energy liability insurance (whether the INSURED is unnamed in such contract and whether or not it is legally enforceable by the INSURED) issued by the Nuclear Insurance Association of Canada or any other group or pool of Insurers or would be an INSURED under any policy but for its termination upon exhaustion of its limits of liability, or

(b) resulting directly or indirectly from the nuclear energy hazard arising from:

(i) the ownership, maintenance, operation or use of any nuclear facility by or on behalf of the INSURED;

(ii) the furnishing by an INSURED of services, materials, parts or equipment in connection with the planning, construction, maintenance, operation or use of any nuclear facility; and

(iii) the transportation, consumption, possession, handling, disposal, or use of radioactive materials (other than radioisotopes away from a nuclear facility sold, handled, used or distributed by an INSURED.

Definitions

(1) The term 'nuclear energy hazards' means the radioactive, toxic, explosive, or other hazardous properties of radioactive material;

(2) The term 'radioactive material' means uranium, thorium, plutonium, neptunium, their respective derivatives and compounds, radioactive isotopes of other elements and any other substances that the Atomic Energy Control Board may, by regulation, designate as being prescribed substances capable of releasing atomic energy, or as being requisite for the production, use or application of atomic energy;

(3) The term 'nuclear facility' means:

(a) any apparatus designed or used to sustain nuclear fission in a self-supporting chain reaction or to contain a critical mass of plutonium, thorium and uranium or any one or more of them;

(b) any equipment or device designed or used for (i) separating the isotopes of plutonium, thorium and uranium, or any one or more of them, (ii) processing or utilizing spent fuel, or (iii) handling, processing or packaging waste;

(c) any equipment or device used for the processing, fabrication or alloying of plutonium, thorium and uranium or any one or more of them if at any time the total amount of such material in the custody of THE INSURED at the premises where such equipment or device is located consists of or contains more than 25 grams of plutonium or uranium 233 or any combination thereof, or more than 250 grams of uranium 235;

(d) any structure, basin, excavation, premises or place prepared or used for the storage or disposal of waste radioactive material;

and includes the site on which any of the foregoing is located, together with all operations conducted thereon and all premises used for such operations.

(4) With respect to property, loss of use of such property shall be deemed to be damage to or destruction of property.